There IS a Planet B

Implementing the Greatest Transformation
in Human History

Sailesh Rao

Cover Concept: Carl LeBlond of Logic to Magic

Cover Design: Mark Hurst of The Chase

ISBN: 979-8-89686-672-5

CopyLeft 2025.

No rights reserved.

Any part of this publication may be reproduced or transmitted in any form or by any means without permission.

There IS a Planet B

Implementing the Greatest Transformation
in Human History

Sailesh Rao

*"The world is not such a terrible place,
as long as you look at it wrong."*
– Glen Merzer

Table of Contents

Acknowledgements	vii
Prologue: How Scientists Gain Public Trust	ix
1. The World Turned Upside Down	2
2. The Greatest Transformation in Human History	12
3. The Four Aspects of Sustainability	18
4. The Four Portals of Veganism	24
5. House on Fire	30
6. Our Planetary Home on Fire	34
7. A Critical Look at the UN IPCC's Emissions Accounting	42
8. The Worst Planetary Boundary Transgression	56
9. Nature the Perfect System Design	60
10. How Humans Belong in Nature	64
11. Setting Context in Time	68
12. The Quaternary Ice Age	74
13. How did we Heat the Climate?	78
14. How we Use The Earth	82
15. The VEGAN Rewilding Solution	88
16. The Animal Agriculture Position Paper	92
17. The Climate Bathtub Model	98
18. The Need for a Spiritual and Cultural Transformation	102
19. From Climate Heating to Climate Healing	104
20. The Core Values Transformation	108
21. The Goals Transformation	112
22. The Self Transformation	120
23. Busting The Protein Myth	126
24. Overcoming the Dairy Deception	132
25. Gaining Independence from Colonialism 2.0	138
26. The Oxford Union Debate	142
27. Codes for a Healthy Earth	150
28. The Civilizational Transformation	158
29. The Money Game Transformation	174
30. How YOU Can Help	180
Epilogue: The Hero's Marathon Journey	182

Acknowledgements

This book was born when I attended the Oxford Literary Festival, 2024, as a guest of Marlene Watson-Tara and Bill Tara. The festival director, Sally Dunsmore, invited me to premiere a book at the Oxford Literary Festival, 2025 and I readily accepted. When she asked me for the title of the book, I blurted out, "There IS a Planet B," a twist on the much quoted title of Mike Berners-Lee's book of facts and figures, "There is NO Planet B".

This book is the outgrowth of a presentation that I've been giving over the years called, "The Greatest Transformation in Human History". The presentation never remained the same as I almost always found something to improve based on the feedback from the audience. I shall remain forever grateful to all those who contributed to this book so directly.

I'm grateful to Marlene Watson-Tara, Bill Tara and Sally Dunsmore for the inspiration to write this book. I'm grateful to the 600 plus presenters and the thousands of participants who have contributed to robust discussions during the 20 Vegan Convergence Of the Peoples (V-COPs) since 2018. They have shaped the narratives in this book immensely.

I wrote this book at home in Phoenix, Arizona, at my cousin's place in Clayton, California, at my other cousin's place in Kansas City, Missouri, at Sarah and Uday's place in Iowa City, Iowa, at JoAnn

There IS a Planet B

and Joe's place in LeCompton, Kansas, at my IIT classmate's place in Mumbai, India, at another cousin's place in Bengaluru, Karnataka, at my sister's place in Mysuru, Karnataka, at yet another cousin's place in Bengaluru, Karnataka, and at Sadhana Forest in Auroville, Tamilnadu, India.

I'm grateful to my family, my cousin, Seetha Suryaprasad, my other cousin, Umesh Kunigehalli and Kalpana Lalgudi, Sarah Vigmostad and Uday Kumar, JoAnn and Joe Farb, Preeta Kannan and Kannan Iyer, Malathi Mohan and Mohan Das, Sudha and Vasanth Kukillaya, Nischita and Uday Chandraghatgi, Yorit and Aviram Rozin and over 100 long-term volunteers at Sadhana Forest, for nourishing me with their love, thought-provoking conversations and healthy Vegan food while I learned from them and wrote this book.

I'm eternally grateful to the members of the core team at Climate Healers for their love and support, for sharing my dream of a Vegan world by 2026 and for making invaluable contributions over the years, helping to shape every single chapter of this book: Alex, Alison, Amit, Ann, Carl, Dakota, Dani, Deborah, Debra, Gabriele, Gerard, Giva, Jamen, Jim, Kelly, Ken, Krish, Lisa, Liz, Maggie, Marco, Paige, Pareen, Paul, Ray, Rebecca, Sarah, Shankar, Stacey, Suzanne, Tami and Vega, the Cow and Climate Healer and her Veguitas.

Finally, I'm especially indebted to Suzanne Ashley-King for her beautiful layout artistry, to Carl LeBlond of Logic to Magic for the cover concept and to Mark Hurst of The Chase for the cover design.

Prologue: How Scientists Gain Public Trust

"Trust has two dimensions:
competence and integrity."
– Simon Sinek

The villagers of Hadagori in the state of Odisha, India, were perplexed. They knew me as a city dweller and they wanted to question me about what was bothering them.

The villagers had started trading with a middleman who was buying vegetables from them to sell in Bhubaneshwar, the capital city of Odisha. The middleman had specific requirements on the vegetables that he would purchase. They had to be clean and without any holes in them. Ugly vegetables had to be picked out. The villagers had to use pesticides to ensure that worms did not feast on the vegetables as they were grown.

The villagers knew that there was something wrong with what they were doing to grow vegetables for the middleman. They planted them far away from the patches where they were planting organic vegetables for their own consumption.

They asked me the question that was bothering them: *"Why do people in the cities want to eat vegetables that even the worms don't want to eat?"*

There IS a Planet B

I was stunned. I later thought about it and realized that as buyers in the city, we don't load our shopping carts with ugly produce or vegetables containing worm holes. This sends a message to the middleman that such produce cannot be sold.

He then figures out how to eliminate such produce from his supply chain and our chemically drenched agricultural processes result. Of course, he legitimately conceals what he is asking the villagers to do from the buyers so that he can continue to grow his business.

The Power to Take Remedial Action

This is how through our consumer choices, we are unwittingly facilitating our own food poisoning. Likewise, through our lifestyle choices, we are also unwittingly participating in exacerbating the nature and climate crisis that is threatening mass extinction, including our own. When we realize this, it gives us the power to take remedial action.

Dr. James Hansen is known as the "grandfather of global warming". He has done pioneering research work teasing out the relationships between human activities and the ongoing nature and climate crisis. In his book, *Storms of my Grandchildren*[1], he wrote,

> *"There is a social matter that contributes equally to the crisis: **government greenwash**. I was startled, while plotting data, to see the vast disparity between government words and reality. Greenwashing, expressing concern about global warming and the environment while taking no actions to actually stabilize climate or preserve the environment, is prevalent in the United States and other countries, even those presumed to be the "greenest."*
>
> *The tragedy is that the actions needed to stabilize climate, which I will describe, are not only feasible but provide additional benefits as well. How can it be that necessary actions are not taken? It is easy to suggest explanations—the power of special interests on our governments, the short election cycles that diminish concern about long-term consequences—but I will leave that for the reader to assess, based on the facts that I will present."*

But what if governments are like the vegetable middleman and climate scientists are like the villagers in the Odisha story? Everyone knows deep down that we need to urgently change how we relate to other life forms on the planet if we are to tackle the nature and climate crisis seriously. The actions needed to stabilize climate, if one were to honestly examine the data, begins with immediate lifestyle and cultural changes, followed by gradual changes to the fossil fuel based energy infrastructure.

The Greenwashing of Lifestyle Issues

But climate scientists like Dr. Hansen are not talking about immediate lifestyle and cultural changes, preferring instead to speak about fossil fuel use. Dr. Hansen did say 15 years ago that animal agriculture is a leading cause of climate change[2], but he has not spoken much about it since.

Throughout the 2024 State of the Climate Report[3], authored by 12 prominent scientists, there were 15 references to fossil fuels and zero references to animal agriculture. In the concluding section, the authors even wrote, *"Rapidly phasing down fossil fuel use should be a top priority."*

From a systems engineering perspective, we know that **it would be an unmitigated global disaster to rapidly phase down fossil fuel use without first shutting down animal agriculture and rewilding the planet**. This is due to what Dr. Hansen has referred to as the Faustian bargain of fossil fuels[4].

When prominent scientists don't adopt the necessary lifestyle and cultural changes and don't talk about the urgent need for them, they behave similarly to greenwashing governments. This **elite greenwashing** of important lifestyle and cultural issues is a social matter that exacerbates the nature and climate crisis.

The actions that we need to take, like dropping animal products from our diets, consuming less, exercising more, etc., are not only feasible, but provide additional benefits for us all. Then why are climate scientists not adopting them or speaking up about them?

There IS a Planet B

Perhaps the brunt of the impacts from the nature and climate crisis falls not on the climate scientists, but on the poor, the marginalized and the indigenous communities, and most acutely on the flora and fauna of the world. Perhaps the climate scientists don't want to take unpopular stands, upsetting their families, friends and grant funders. Perhaps they are habituated or even addicted to their traditions, lifestyles and culture.

I will leave that for the reader to assess based on the facts that I present in this book, but what if the greenwashing governments are just mirroring greenwashing scientists and by extension, greenwashing citizens?

We know that there is **citizen greenwash** on lifestyle issues, since almost everyone claims not to want to hurt animals unnecessarily, while the data shows that we are eating animal foods and causing unnecessary suffering to animals.

Perhaps we have been misled by school science teachers into thinking that protein is only found in animal foods and calcium is only found in dairy foods. Perhaps we too are habituated or even addicted to our traditions, lifestyles and culture.

How we Can Solve This Together

This book is an invitation for prominent scientists like Dr. Hansen to engage with us on the necessary lifestyle and cultural changes and speak out about them openly in order to start the conversation with the general public. This will create the momentum for mass grassroots action so that Dr. Hansen's grandchildren, someday in the future, will look back and say, *"Opa understood what was happening, and he made it clear."*

This book is also an invitation for the reader to have faith that those who are still ensnared in choices of personal and planetary destruction are just waiting for the right moment to break free. It is an invitation to have faith in humanity, not hope, because hope comes from a position of fear, while faith comes from a position of love.

Have faith that we already have all the tools and technologies to ensure that every human being is able to implement the necessary lifestyle and cultural changes in every corner of the globe. Have faith that causing unnecessary suffering to innocent animals is already widely recognized as a moral wrong.

Have faith that the lie of endless economic growth, of food systems rooted in suffering, is coming to an end – not through force, but through an irrepressible wave of love, kindness and compassion in action.

Have faith that we are living in a world of abundance, not scarcity.

Have faith that the necessary global transformation is already happening everywhere and that the path to personal and planetary healing is not only possible, but actually inevitable.

This is not a question of whether, but how fast.

Engaging with Naysayers

If anyone tells you that it is already too late and our species is doomed to near-term extinction anyway and so why bother changing, please ask them, *"How would you know?"*

Life is a complex, emergent, nonlinear process which is impossible to model accurately. As a systems engineer, I used to do modeling for a living and I know that even a single significant nonlinearity makes it difficult to get the model to predict certain aspects of reality reliably. For instance, it is notoriously difficult to predict the ***tipping point*** for a nonlinear feedback phenomenon, a point of no return beyond which it is impossible to reverse the condition.

The nature and climate crisis is governed by thousands of significant nonlinearities with both positive and negative feedback loops. For such a complex reality, the best we can do is to estimate the ranges for the tipping points using our models. We have to acknowledge with humility that we cannot predict the exact tipping points reliably.

There IS a Planet B

In this case, the only way to know if we have crossed a tipping point is to initiate a forcing in the opposing direction, i.e., actually reduce the greenhouse gas concentrations in the atmosphere, plant native trees in a region, etc., and verify whether the system is responding in the right direction, i.e., whether the Arctic sea ice cover begins increasing year on year, native wild animals return to the region, etc.

Even if the system does not respond in the right direction, we would still be unsure if we have crossed the tipping point or not. We can then increase the forcing in the opposing direction and see if it responds. This is why as long as photosynthesis is still working and there is sufficient oxygen in the atmosphere, we will have air to breathe, food to eat, a life to live, a lot to love and plenty of work to do. We certainly won't have time for hand-wringing and such tales of doom.

Please read on...

---

1 The quoted passages are taken from the Preface of J.E. Hansen, Storms of My Grandchildren: The Truth About the Coming Climate Catastrophe and Our Last Chance to Save Humanity," Bloomsbury USA, 2010. Link https://www.bloomsbury.com/us/storms-of-my-grandchildren-9781608195022/ accessed on March 4, 2025.

2 In an interview with Supreme Master TV in 2010, Dr. Hansen asked humanity to "Be veg, go green and save the planet," https://www.youtube.com/watch?v=zfbxeAFk8mY

3 The State of the Climate Report is an annual report published by Oxford Academic BioSciences and authored by William Ripple, et al. in 2024. I wrote a letter to the Editor of BioSciences requesting a correction to the Report regarding its blackout on animal agricultural issues, but as of the publication of this book, I have not received a response. Link https://academic.oup.com/bioscience/article/74/12/812/7808595?login=false accessed on March 4, 2025.

[4] The Faustian bargain of fossil fuels is that burning fossil fuels releases cooling gases that mask the heating effects of the warming gases. Hansen et. al, *Climate Forcing Growth Rates: Doubling Down On Our Faustian Bargain*, Environmental Research Letters, Vol 8, No 1, 2013. https://iopscience.iop.org/article/10.1088/1748-9326/8/1/011006/meta link accessed on March 8, 2025.

1. The World Turned Upside Down

"You never change things by fighting the existing reality. To change something build a new model that makes the existing model obsolete."
– Buckminster Fuller

We have been looking at everything the wrong way around. If you subscribe to the notion that, *There is NO Planet B*[1], then you might believe that we have no choice but to continue spiraling downward on our current path of self destruction and ecological annihilation.

That is not true. **There IS a Planet B**.

Planet B is the current world of Planet A turned upside down. To be more precise, on Planet A, we have been looking at reality upside down.

Planet A is a Company Town

On Planet A, with its climate *heating* phase of our civilization, built on a foundation of false myths, the media apparatus is greased to rapidly spread lies in order to keep the global human population addicted, sick and compliant. Therefore, it is not surprising that lies spread much faster than the truth and become difficult to dislodge once they get stuck in our minds.

There IS a Planet B

Planet A is actually a company town.

The company is primarily in the business of addiction. It is incredibly profitable and makes a few people fabulously wealthy. The company has numerous global subsidiaries and most of them are either in the business of addiction or in providing the tools and technologies needed for the business of addiction. In its early stages, the company ran an opium business[2] and then it diversified into other addictive substances like tea, coffee, sugar, salt, tobacco, alcohol, prescription drugs and technologies like social media. It operated two huge opium factories in India in the 19th century and even went to war with China to force easy access to users there. One of these factories in Ghazipur, Uttar Pradesh, India, is still operating today, but it is now supplying opium for the pharmaceutical industry. Its business is still addiction, but with a more respectable veneer.

The Consequences of Addiction

Addiction has its consequences.

In the seventh decade of my existence on Planet A, every day feels like a tremendous bonus. Little did I know when I embarked on my technical career as a systems engineer that during the later years of my life, I would be championing a spiritual and cultural transformation of ourselves and our civilization to stabilize life-support systems on our planet.

I was born in India a dozen years after India gained political independence from direct British rule. At birth, my life expectancy was 42 years, an improvement over 36 years[3], the life expectancy of a child born at the time of independence in 1947. The improvement in life expectancy was mainly due to better maternal care, which reduced infant mortality rates in India.

I came to the US to pursue my graduate studies at the State University of New York in Stony Brook, Long Island in 1981. In the first letter that I wrote to my parents from the US, I confessed that I had started smoking during my college days. I had been hiding that

3

fact from my parents for a couple of years and I was feeling guilty about it.

A few weeks later, I received a six-page letter from my mother handwritten in my native language, Kannada. In it, she revealed to me for the first time that I was a twin, that my twin died at childbirth, and that I was a sickly infant with breathing difficulties. She wrote that she had to massage my chest with essential oils for the first two years of my life in order to help me breathe. Then, she pleaded with me to quit smoking.

From that day on, she made it her mission to persuade me to quit smoking. In every letter and in every phone call between us, she inquired about the progress of my smoking cessation efforts.

I tried every smoking cessation product on the market, nicotine gums, nicotine patches, you name it. But no matter how much I tried, I couldn't quit smoking.

Then, in 1996, my mother was diagnosed with heart disease. She was already suffering from diabetes and the cardiologist told us that she would likely not survive a surgical procedure to fix her heart.

I was devastated. I went to a hypnotherapist and begged him to help me quit smoking. His therapy succeeded for about two weeks until a particularly stressful event at work caused me to take up smoking again.

On February 20, 1997, my mother died in her sleep, literally of a broken heart. I was numb with grief that I couldn't do the only thing that she ever really asked me to do when she was alive.

I begged our family doctor for help. He said that a new anti-depressant had just come on the market which could be used to help people quit smoking. You just had to take two tablets every day and continue smoking as usual. For about 40% of the population, it had the effect of eliminating the urge to smoke, sometimes permanently. I filled in a 100 tablet prescription for the anti-depressant and started the process. I took two tablets on the

4

There IS a Planet B

first day and I didn't feel like smoking at all. The next day, I didn't even feel like taking the tablets.

I knew I had kicked my smoking habit for good. I haven't smoked ever since.

Lessons from a Mother

One of my greatest regrets until today is that I waited until my mother died before overcoming my addiction. Now, I know that I always had it in me to kick that addiction if only I had summoned the discipline to do so when my mother was still alive.

Nothing worthwhile ever gets done without discipline. We are not going to solve the nature and climate crisis without discipline. We are not going to end world hunger without discipline. We are not going to achieve world peace without discipline.

Since then, I have put this lesson from my mother to good use. When I discovered that drinking coffee and alcohol was having a detrimental impact on my physical health in 2005, I quit coffee and alcohol instantly.

When I discovered that consuming dairy is detrimental to our planetary health in 2008, I ditched dairy products and adopted a vegan ethic and lifestyle instantly.

And above all, the greatest lesson I learned from my mother is that we must summon the discipline to overcome our addictions and create the world of Planet B before our Mother Earth dies.

There are no do-overs this time around.

The Four Foundational Myths on Planet A

The good news is that the foundation of falsehoods on Planet A is already crumbling before our eyes in the light of the truth.

There are four foundational myths baked into Planet A:

5

1. Protein is only found in animal foods.
2. Calcium is only found in dairy foods.
3. Human well-being necessitates endless economic growth.
4. Fossil fuel use is the leading cause of climate change.

A myth is a lie wrapped in a truth. For instance, it is true that protein is found in animal foods, but protein is also found in all plant foods. However, when science teachers repeatedly drill into our heads that protein is found in animal foods, we begin to believe that protein is ONLY found in animal foods.

Hence the question that I have repeatedly answered, *"Where do you get your protein?"*

Those who knowingly promote these myths, even to innocent children, must truly have developed hearts of stone as the carnage they unleash results in tens of millions of deaths each year. Perhaps they believe that they are "managing" the human population just like they are managing wildlife populations with guns, glue traps and culling expeditions.

The average global diet, with the meat and dairy heavy Standard American Diet (SAD) as its ideal, is responsible for tens of millions of deaths worldwide each year, both from calorific excess and global hunger. Hunger and hunger-related causes alone kill about 9 million people[4] each year.

The global food system procures almost six times as much food as humanity needs from the planet and yet manages to starve 800 million people while it renders 95% of Americans[5] in the richest country in the world chronically deficient in an essential macronutrient, fiber, not to mention numerous micronutrients.

This is because the food system on Planet A is optimized to ensure that the most exotic food is available to the richest person at any time. It is not optimized to ensure that all humans have access to healthful foods providing adequate nutrition.

The endless economic growth paradigm is so woven into the fabric of society on Planet A that no one blinks an eyelid even when for

There IS a Planet B

example, Norway, a country whose Country Overshoot Day[6] fell on April 16 in 2024, boldly proclaims in its Climate Action Plan[7] that it is **a plan to cut emissions, not economic growth.**

Every year, a Country Overshoot Day marks the date when the planet's annual ecological capacity budget would be used up if everyone on Earth lived at the same level of consumption as the residents of that particular country. For the remainder of the year, humanity would be living on the backs of future generations.

It is amazing how everyone is programmed on Planet A to believe that striving for economic growth even when a nation has already grown three, four or even ten times more than what the planet can support is as normal as eating food, drinking water, breathing air or pursuing happiness.

Finally, at the turn of this century, the fourth foundational myth on fossil fuel use was put in place in order to prop up the other three. In order for mainstream climate spokespeople to proclaim that fossil fuel combustion is the leading cause of climate change, the United Nations (UN) Intergovernmental Panel on Climate Change (IPCC) politicized emissions accounting[8] mainly by:
- undercounting land use change emissions by a factor of 3;
- undervaluing methane emissions by a factor of 3;
- not counting the cooling effects of fossil fuel combustion; and
- ignoring the Carbon Opportunity Cost of the land used for animal husbandry.

Climate change related deaths already total an estimated 5 million per year[9] worldwide. If the world implements the recommendation in the State of the Climate Report: 2024[10] to "rapidly phase down fossil fuel use as a top priority," it would **rapidly increase the anthropogenic global warming on the planet by 50-87% in a matter of weeks.**

The Stone-Hearted and the Clay-Footed

The mind boggles as to how much that might increase climate related deaths into the tens or even hundreds of millions per year

amidst planetary destruction. Only the stone-hearted can contemplate this catastrophe with equanimity.

Enabling the stone-hearted are the clay-footed who knowingly stay silent as the stone-hearted tell lies and commit atrocities in order to maintain the status quo that is keeping them in power, regardless of the consequences. The clay-footed don't have the courage to call out the stone-hearted and tell the truth.

When we know better, shouldn't we do better?

Between the stone-hearted and the clay-footed, it is the behavior of the clay-footed that is truly heart-breaking on Planet A. As Elie Wiesel said, *"What hurts the victim the most is not the cruelty of the oppressor, but the silence of the bystander."*

It is the lack of courage of the forces of light that lets the forces of darkness run rampant. Lack of courage is the single greatest obstacle to social and ecological progress at the moment.

The Four Foundational Truths on Planet B

The climate *healing* civilization on Planet B is built on a foundation of openness, integrity and the four Foundational Truths:
  1. All plant foods have all amino acids and have complete protein;
  2. If you eat enough calories on a natural diet, you will get enough calcium;
  3. Human well-being necessitates a sense of purpose larger than ourselves; and
  4. Animal husbandry is the leading cause of climate change, while stopping fossil fuel use now is suicidal.

A system implemented on these four foundational truths unlocks the underlying abundance on Planet B so that
  • there is zero hunger because access to healthy nutritious food is a fundamental human right.
  • there is no poverty because everyone's basic needs are met by the community.
  • good health and well being is common because healthy habits are normalized in childhood itself.
  • quality education is freely available to everyone.

There IS a Planet B

- honesty and humility are common traits among humans in relation to each other and in relation to the last remaining wild creatures that we still share the Earth with.

The remaining chapters are an exploration of how we can accelerate this Greatest Transformation in Human History from the climate *heating* phase of our civilization on Planet A to its climate *healing* phase on Planet B.

---

[1] *There is NO Planet B: A Handbook for the Make or Break Years* is a book of facts and figures authored by Mike Berners-Lee and published by Cambridge University Press in 2021. Link http://theresnoplanetb.net/ accessed on March 4, 2025.

[2] *Smoke and Ashes: Opium's Hidden Histories* is a searing non-fiction account about how the colonial opium trade morphed into trade in other addictive substances like tea, coffee, sugar, cocoa and finally, pharmaceuticals and the modern day opioid crisis. It is written by Amitav Ghosh and was published by Farrar, Strauss and Giroux in 2024. Link https://www.waterstones.com/book/smoke-and-ashes/amitav-ghosh/9781529349269 accessed on March 4, 2025.

[3] *Mortality Trends and Patterns in India: Historical and Contemporary Perspectives* by K. Navaneetham and C. S. Krishnakumar is Chapter 11 of the book, *Shaping India: economic change in historical perspective*, ed by D. Narayana and Raman Mahadevan, published by Routledge, New Delhi, 2011. This chapter analyzes the life expectancy of a child born in India at various times and the causes of mortality. It makes for fascinating reading. Link https://www.researchgate.net/publication/273461759_Mortality_Trends_and_Patterns_in_India_Historical_and_Contemporary_Perspectives accessed on March 4, 2025.

[4] According to the United Nations, approximately 700-800 million people are hungry and 9 million people die from hunger and hunger-related causes each year. This includes malnutrition, which weakens the immune system and makes people more vulnerable to diseases.

A significant portion of these deaths are due to conditions in developing countries, where access to food, healthcare, and clean water can be scarce. The UN's Food and Agriculture Organization (FAO) and World Health Organization (WHO) regularly monitor these figures. The latest WHO/UN FAO report on *The State of Food Security and Nutrition in the World*. Link https://www.who.int/publications/m/item/the-state-of-food-security-and-nutrition-in-the-world-2024 accessed on March 4, 2025.

[5] About 95% of Americans do not consume enough fiber. The recommended daily intake of fiber is around 25 grams for adult women and 38 grams for adult men, according to the Dietary Guidelines for Americans. However, studies consistently show that the average American only consumes about 15 grams of fiber per day. For more information, please see D. Quagliani, P. Felt-Gunderson, *Closing America's Fiber Intake Gap: Communication Strategies from a Food and Fiber Summit*, American College of Lifestyle Medicine, Vol 11, Issue 1, 2015. Link https://doi.org/10.1177/1559827615588079 accessed on March 4, 2025.

[6] A Country's Overshoot Day is the day on which Earth Overshoot Day would fall if everyone in the world consumed like the people in that country. Country Overshoot Days are published on Jan 1. of each year and the 2024 data can be found here: https://overshoot.footprintnetwork.org/content/uploads/2024/01/Country-Overshoot-Days-2024.pdf, accessed March 4, 2025.

[7] Norway's Climate Action Plan for 2021-2030 can be accessed here: https://www.regjeringen.no/en/dokumenter/meld.-st.-13-20202021/id282740 5/?ch=1 accessed on March 4, 2025.

[8] Please see the explanation in https://climatehealers.org/blog/a-critical-look-at-un-ipccs-emissions-accounting/, which is largely reproduced in Chapter 7 of this book.

There IS a Planet B

---

[9] A Monash University study in 2021, linked 5 million extra deaths each year to abnormal temperatures, i.e., climate related causes. It was published in the Lancet: Qi Zhao, et al., *Global, regional, and national burden of mortality associated with non-optimal ambient temperatures from 2000 to 2019: a three-stage modelling study*, The Lancet Planetary Health, vol 5, issue 7, July 2021. Link https://www.thelancet.com/journals/lanplh/article/PIIS2542-5196(21)00081-4/fulltext accessed on March 4, 2025.

[10] The State of the Climate Report is an annual report published by Oxford Academic BioSciences and authored by William Ripple, et al. in 2024. Link https://academic.oup.com/bioscience/article/74/12/812/7808595?login=false accessed on March 4, 2025.

2. The Greatest Transformation in Human History

"The greatest human quest is to know what one must do
to become a human being."
– Immanuel Kant

The company town of Planet A is in crisis. Its human systems are in crisis because they are embedded within a planetary crisis that is ecological in nature. These systems are designed to take from the planet, not give to the planet and this is the root cause of the ecological crisis it faces today.

As James Lovelock pointed out[1], *"If the earth improves as a result of human presence, then we will flourish. If it does not, then we will perish."*

On Planet A, the earth is not improving as a result of human presence. Far from it. Mother Earth is literally dying in front of our eyes making it well past time that we overcome our addictions that are killing her.

Although I'm an environmentalist by occupation, I'm a systems engineer by profession. I received my doctoral training at the Information Systems Laboratory at Stanford University and I spent the first two decades of my professional career working on internet communication systems. During this period, I developed the

There IS a Planet B

protocol for transforming early analog internet connections into more robust digital connections while accelerating their speed tenfold. Today, over a billion internet connections still employ that 1000BASE-T Gigabit Ethernet protocol worldwide.

After watching former Vice President Al Gore's "global warming" slide presentation[2] on TV in 2005, I switched careers and became deeply immersed, full time, in solving the nature and climate crisis facing humanity. I founded Climate Healers in 2007 to help bring systems thinking to the fore in dealing with this crisis.

A crisis is a powerful point of transformation and the nature and climate crisis is a powerful point of the Greatest Transformation in Human History from the climate *heating* phase on Planet A to the climate *healing* phase of our civilization on Planet B. This transformation is more significant than the Industrial revolution, the Agricultural revolution, the Scientific revolution, and even the discovery of fire, for it is the discovery of the spiritual fire within humanity, a revolution of what it means to be human.

Our tag line at Climate Healers is:
Transform Yourself
Transform our World

We use the word "transform" as opposed to "change" because as Tracey Martin observed[3], *"Change is inevitable, while transformation is intentional."*

The Greatest Transformation in Human History is founded on the fastest growing social justice movement in the world today, the Vegan movement, already boasting over 200 million conscious adherents[4] worldwide. In my case, I went Vegan suddenly when I was visiting 200 acres of protected land in the village of Karech in Rajasthan, India, in 2008.

My Vegan Transformation

There was a stone fence protecting that land and to the left of the fence, I could see old, former dairy cows walking around eating

anything new that was growing on the ground. To the right of the fence within the protected land, I could see a lush green forest.

I immediately realized that my consumption of dairy products was causing the forest to die. As an ethical lactovegetarian at that time, I was consuming dairy products, but I also did not want the cows to be killed when they couldn't produce milk as they got older.

The people of Karech were doing just what I wanted dairy farmers to do. They were letting their dairy cows live out their lives after they were finished with their milk-producing years.

I could now connect the dots on the consequences of that decision.

I realized that in an ethical lactovegetarian world, for each milk producing animal, there would be a male counterpart, which would double the number of animals. For cows, the milk production occurs over roughly one-quarter of their lifespan in the modern world and therefore, if we allow the cows to live out their lives, that would further quadruple the total number of animals in the global dairy herd.

Each milk-producing cow was lactating because she recently gave birth to a calf. If we allow the calf to drink half the milk from the mother cow, then that would further double the number of animals needed to produce the same amount of milk.

Putting it all together, in an ethical lactovegetarian world[5], the dairy herd would have to increase by a factor of 2×4×2=16, which would bring the global dairy herd size to over 12 billion cows, buffaloes and other milk producing animals in the food system.

This is simply impossible to sustain on planet Earth.

Once I realized this, I went Vegan on the spot in the village of Karech, Rajasthan, India, in 2008.

Now there is little doubt that shutting down animal husbandry is by far, the single most effective step we can all take to address the nature and climate crisis and transform Planet A into Planet B.

There IS a Planet B

However, you won't hear this "sustainability secret" from most climate spokespeople, who are reluctant to admit the true impact of animal husbandry on the ecological health of the planet. A decade ago, I witnessed this reluctance first hand.

The Origins of Scientific Reticence

The American Geophysical Union (AGU) Fall meeting is the largest annual gathering of climate scientists in the world. When I attended the meeting in 2015, I spoke to numerous climate scientists who privately agreed with me that the cessation of animal husbandry can literally reverse climate change, but they thought that would never happen.

During the banquet dinner at the meeting, I understood why they thought that would never happen. I had to wait for the chef to prepare a special vegan meal of pasta and tomato sauce as the main course was a steak dish and the alternate main course was lasagna loaded with cheese.

Many climate scientists are too scared to tell the truth on this topic and prefer to go along with the false framing of climate change as mainly a fossil fuel issue. One prominent scientist who works at an elite university and contributes to the UN IPCC told me that he cannot tell the truth on this topic and keep his job at the same time.

As Upton Sinclair put it so bluntly, *"It's difficult to get a man to understand something when his salary depends upon his not understanding it."*

Outing the Sustainability Secret

Despite this reluctance, the sustainability secret is finally out. All of the events I attended during the Los Angeles Climate Week of 2024 were completely vegan. Twenty three New York city restaurants totaling 38 locations were highlighting vegan dishes[6] during New York Climate Week, 2024.

The New York Times served a vegan lunch to Jane Goodall and other guests at its Climate Forward event during that week. The New York Times even published an opinion piece a few years ago stating, *"Vegans are right about ethics and the environment. If you won't join them, at least respect their effort to build a sustainable future."*

The Times of India newspaper dated February 1, 2025 quoted the Economic Survey of the Finance Ministry of the Indian Government of 2025 making the case for going Vegan and avoiding meat and dairy consumption. It stated that cutting meat and dairy could reduce emissions[7] by 66%. It also pointed out that 17% of the food is wasted and if food waste were a country, it would be the third largest emitting country in the world!

The true impact of animal husbandry on the environment is now becoming common knowledge.

As the Buddha said, *"Three things cannot be long hidden: the sun, the moon and the truth."*

---

[1] This quote is taken from his interview on BBC Hardtalk: James Lovelock: The Future of Life on Earth. Link https://www.bbc.com/audio/play/w3ct1n5y accessed on March 4, 2025.

[2] The slide presentation that I saw was reproduced in the documentary, An Inconvenient Truth. Link https://en.wikipedia.org/wiki/An_Inconvenient_Truth accessed on March 4, 2025.

[3] Tracey L. Martin is a transformational coach and she said these exact words to me during a meeting in Phoenix, AZ. You can find more about Tracey and contact her here: https://www.traceylmartin.com/ accessed on March 4, 2025.

There IS a Planet B

---

[4] The exact number of vegans in the world is difficult to pin down. The largest online survey done by Statista with over 181,000 respondents aged 18-64 between Jan to Dec 2024 estimated that around 280 million people around the world are vegan. However, I'm aware that the number of vegans in India has been overestimated in the survey as people in India tend to confuse vegetarianism with veganism. Hence, my estimate is a bit lower. Link https://www.statista.com/statistics/1280066/global-country-ranking-vegan-share/ accessed on March 4, 2025.

[5] Please see S. Rao and L. Barca, *The Ethical Vegetarian Myth,* which was published as a chapter in *The Humane Hoax: Essays Exposing the Myth of Happy Meat, Humane Dairy and Ethical Eggs*, ed. by Hope Bohanec, Lantern Press, 2023. Links https://climatehealers.org/the-science/ethical-vegetarian-myth/ and https://lanternpm.org/book/the-humane-hoax/ accessed on March 4, 2025.

[6] New York Climate Week happens in September of each year to raise awareness of the need for climate action. Live Kindly reported that 23 restaurants totaling 38 locations went Vegan for the week here: Link https://www.livekindly.com/nyc-restaurants-vegan-dishes-eat-for-climate-week/, accessed March 4, 2025.

[7] The Economic Survey of India is published by the Chief Economic Advisor of India, Dr. V. Anantha Nageswaran and the quotes are from Chapter 10 of the Economic Survey: Climate and Environment: Adaptation Matters. Link https://www.indiabudget.gov.in/economicsurvey/doc/eschapter/echap10.pdf accessed March 4, 2025.

3. The Four Aspects of Sustainability

"Sustainability is treating ourselves and our environment as if we are to live on this earth forever."
– Arron Wood

In 1987, the United Nations Brundtland Commission defined sustainability[1] as *"meeting the needs of the present without compromising the ability of future generations to meet their own needs."*

Although this definition focused just on the needs of humanity, it did capture the infinite nature of the sustainability paradigms that we are transitioning towards on Planet B, from the winner-takes-all, finite nature of the unsustainable paradigms we employ on Planet A.

Sustainability is the *infinite* game, not a *finite* game[2], since our aim is to live on this planet forever. The purpose of the infinite game is to continue the game forever, with players rotating in and out of the game as needed. In contrast, the purpose of a finite game is to win the game. Please note that while there are infinitely many finite games, there is only one infinite game.

Until we change our purpose as a species so that we can be playing this infinite game, it is unlikely that we will ever successfully achieve sustainability.

There IS a Planet B

In the climate *heating* phase of our civilization on Planet A, we have been mainly playing finite games while depleting the ecological wealth of our planet and heating the planet. We determine winners and losers in the games and reward the winners at the expense of the losers.

In the climate *healing* phase of our civilization on Planet B, we will mainly be playing the infinite game while striving to restore the ecological wealth of our planet and cooling the planet.

The transformation from the climate *heating* phase to the climate *healing* phase requires addressing the four main aspects of sustainability – ecological, economic, ethical and educational.

Ecologically, we need to restore ecosystems and stabilize the life-support systems of the planet so that life on earth can thrive forever

Economically, we need to ensure that the present needs of humanity are met.

Ethically, we need to ensure that there is equity and justice for all, regardless of their race, gender, age, ability or species identity.

We have built the world that we were educated to build. Since the world we have built is not sustainable, we need to revamp our education system so that the world we build in the future is indeed sustainable.

Is True Sustainability Feasible?

True sustainability is at the nexus of all four aspects of this transformation.

The question arises as to whether such true sustainability is even feasible, given that there are 8 billion human beings on the planet today.

As an Executive Producer of *Cowspiracy: The Sustainability Secret*[3], a Vegan environmental documentary that was released in 2014, I sent the video link to every board member of the Climate Reality

Project[4], with a request for feedback. The only feedback I received was from a former CEO of the National Wildlife Federation who declared that *"the world will not go vegan until the human population is down to 1 billion."*

He was echoing what others have also told me, that the earth is already overpopulated with humans and some 7 out of every 8 humans will have to die off before we can achieve sustainability. Undoubtedly, such a mindset is at the root of widespread global apathy on the nature and climate crisis and it will continue to drive destructive patterns until we find the courage to confront it.

Fortunately, such a mindset is based on culture, not science. The culture of Planet A assumes that every human is wired to acquire more and more material possessions. The typical American is consuming at a level that is over 5 times more than what the planet can sustainably support. The Country Overshoot Day[5]   for the United States was March 13th in 2024.

A typical American retired executive like my good friend at the Climate Reality Project would be consuming at a level that is double the national average, about 10 times more than what the planet can sustainably support. Then he assumes that every individual's consumption can be capped at his consumption level so that in a sustainable world, there would be at most 8 billion/10 or roughly 1 billion people.

If a billionaire consuming at a level that is 80 times more than what the planet can sustainably support were to make the same calculation, then he would proclaim that the sustainable population of the planet is 100 million human beings or less.

And so it goes.

However, there is no scientific basis for claiming that the planet cannot feed 8 billion human beings, when it is already feeding 7.2 billion human beings and over 40 billion land animals, with the animals eating almost five times as much food as all human beings put together.

There IS a Planet B

There is no problem with implementing a food distribution network to feed 8 billion human beings either as the food system is now feeding over 40 billion land animals an adequate diet and fattening them up for consumption.

Similarly, there is no dearth of shelter or clothing to adequately provide for 8 billion human beings on the planet. However, we humans tend to assume that whatever is being done today will continue to be done forever.

Changing our minds is hard, not just in our dietary preferences, but in order to solve our problems, we need to think differently.

In 1996, as a systems specialist in the standards committee working on internet communications protocols at the Institute for Electrical and Electronics Engineers (IEEE), I was asked to examine why there were significant field returns of the internet devices that were already deployed.

I studied the analog communications protocol in those devices and came to the conclusion that if it were transformed to a digital signal processing protocol, then not only could we make the devices more robust, but we could also increase their speed ten-fold. Given the advances in high speed analog to digital converters and signal processing hardware, such an analog to digital transformation was entirely feasible.

At the next standards meeting, when I made my proposal for such an analog to digital transformation, everyone laughed at me. The chairperson of the committee said to me, *"We are having trouble getting our devices to work and you are talking about cranking up their speed ten-fold. I'll believe it when I see it."*

But the committee was kind enough to let me and my colleagues pursue these ideas further.

By the end of 1996, our startup, Silicon Design Experts, was acquired by Level One Communications and in 1999, Level One was acquired by Intel Corporation for 2.2 billion dollars[6]. The Gigabit Ethernet protocol[7] was standardized by the IEEE in 1999 and by

2003, devices implementing this protocol were on every motherboard at Intel, powering 150 million computers in that year alone.

The internet revolution was well underway.

Likewise, the transformation from the climate *heating* phase on Planet A to the climate *healing* phase on Planet B is entirely feasible today and it is best accomplished sooner, without waiting for 7 billion human beings to die off, or worse yet, actively engineering their accelerated demise on Planet A through hunger, wars and walls.

---

[1] The Brundtland Commission is officially known as the *Report of the World Commission on Environment and Development: Our Common Future* chaired by Gro Brundtland, former Prime Minister of Norway. Link http://www.un-documents.net/our-common-future.pdf accessed on March 4, 2025.

[2] The distinction between finite games and infinite games follows the work of James Carse in *Finite and Infinite Games: A Vision of Life as Play and Possibility*, Free Press, 2013. Link https://jamescarse.com/books/finite-and-infinite-games/ accessed on March 4, 2025.

The entire first chapter of the book is: *"There are at least two kinds of games. One could be called finite, the other infinite. A finite game is played for the purpose of winning, an infinite game for the purpose of continuing the play."*

And the entire last chapter is: *"There is but one infinite game."*

[3] *Cowspiracy: The Sustainability Secret* is a ground-breaking documentary highlighting the environmental impact of animal agriculture, featuring Kip Andersen as protagonist. Link https://cowspiracy.com/ accessed on March 4, 2025.

There IS a Planet B

---

[4] *The Climate Reality Project* is a non-profit organization founded by former Vice President Al Gore in 2006 to raise awareness on the impact of burning fossil fuels on climate change. The Climate Reality Project projects a false framing of climate change since it fails to address the cooling effects of fossil fuel combustion and ignores the impact of animal agriculture on climate change. Vice President Al Gore and his partner David Blood manage *Generation Investment Management*, a $42B fund founded in 2004 to advance a clean energy economy. Links https://climaterealityproject.org/ and https://www.generationim.com/ accessed on March 4, 2025.

[5] A Country's Overshoot Day is the day on which Earth Overshoot Day would fall if everyone in the world consumed like the people in that country. Country Overshoot Days are published on Jan 1. of each year and the 2024 data can be found here: Link https://overshoot.footprintnetwork.org/content/uploads/2024/01/Country-Overshoot-Days-2024.pdf, accessed March 4, 2025.

---

[6] In its largest acquisition ever until that time, Intel Corporation acquired Level One Communications on March 4, 1999 for $2.2B, in stock in an effort to expand Intel's reach into the fast-growing computer network equipment business. Link https://money.cnn.com/1999/03/04/news/intel/ accessed on March 4, 2025.

---

[7] In computer networking, Gigabit Ethernet (GbE or 1 GigE) is the term applied to transmitting Ethernet frames at a rate of a gigabit per second. The most popular variant, 1000BASE-T, is defined by the IEEE 802.3ab standard. For more, please see https://en.wikipedia.org/wiki/Gigabit_Ethernet accessed on March 4, 2025.

4. The Four Portals of Veganism

"It costs us so little to go vegan.
It costs us so much if we don't."
– Gary L. Francione

Veganism is the foundation of the climate *healing* phase on Planet B just as animal husbandry is the foundation of the climate *heating* phase on Planet A.

Veganism is an ancient concept practiced for millennia all across the world. At Climate Healers, we define **VEGAN** through the acronym, **V**itally **E**ngaged **G**uardians of **A**nimals and **N**ature. It captures the social justice aspects of Veganism succinctly and instils the necessary sense of purpose in humanity.

Officially, according to the Vegan Society founded in 1944, the modern definition of Veganism is:

"a philosophy or way of living which seeks to exclude — as far as is possible and practicable — all forms of exploitation of, and cruelty to, animals for food, clothing or any other purpose; and by extension, promotes the development and use of animal-free alternatives for the benefit of animals, humans and the environment. In dietary terms, it denotes the practice of dispensing with all products derived wholly or partly from animals."

There IS a Planet B

As we can see from the definition, Veganism is about moral intention, not moral perfection. It is the next stage in the inexorable progress of civilization towards more justice.

Mahatma Gandhi bent the arc of the moral universe towards justice by getting a majority of people to acknowledge that if colonialism is not wrong, then nothing is wrong. Rev. Martin Luther King, Jr., bent the arc of the moral universe towards justice by getting a majority of people to acknowledge that if racism is not wrong, then nothing is wrong. Today, we have get a majority of people to acknowledge that if causing unnecessary suffering to animals is not wrong, then nothing is wrong.

Veganism in Major Religions

Hindu sages have been practicing Veganism since pre-Vedic times. They went up the foothills of the Himalayas seeking enlightenment and they did not take cows with them. Rabbi Dr. Gabriel Cousens adopted Veganism for spiritual reasons in the 1970s, following the example of these Hindu sages.

In that sense, Veganism is just the modern world acknowledging that our ancestors got it right when they said, *"Ahimsa Paramo Dharma"* or nonviolence is the highest virtue.

Veganism is a practical implementation of compassion for all creation, the foundation of all major religions. Therefore, every single major religion encourages Veganism in the core teachings.

In the Old Testament, sacred to all the Abrahamic faiths, Judaism, Christianity and Islam, *"I give you every seed bearing plant on the face of the whole earth and every tree that has fruit with seed in it. They will be food for you."*

An Islamic Hadith goes, *"A good deed done to an animal is as meritorious as a good deed done to a human, while an act of cruelty done to an animal is as bad as an act of cruelty done to a human."*

25

The founders of the Unity Church of Christianity wrote extensively about Veganism: *"Our food should be full of life in its purity and vigor. There should be no idea of death and decay connected with it in any degree. The vegetable should be fresh and the fruit radiant in its sunny perfection."*

When asked by a reader why their books are not bound in leather, the Unity publishers wrote back that Unity opposes the use of any product that necessitates the taking of life, whether it is a food substance, wearing apparel or general utility.

The Bhagavad Gita says, *"Those who hate no one and are kind to all creatures are dearest to God."*

The Buddha said, *"All beings tremble before violence. All love life. All fear death. See yourself in others. Then, whom can you hurt? What harm can you do?"*

In Jainism, the prohibition of cruelty to all living beings is central to the religion. Lord Mahavira said, *"Do not injure, abuse, oppress, enslave, insult, torment, torture or kill any creature or living being."*

Veganism and Great Thinkers

Great thinkers throughout the ages have associated social justice with our treatment of animals. Pythagoras wrote, *"As long as man continues to be the ruthless destroyer of lower living beings, he will never know health or peace. For as long as men massacre animals, they will kill each other."*

Leo Tolstoy concurred nearly 2500 years later, *"As long as there are slaughterhouses, there will always be battlefields."*

He went on to write:

> *"The immorality of eating animal foods has been recognized by all mankind during all the conscious life of humanity. Why then have people generally not come to acknowledge this law? The answer is that the moral progress of humanity is always slow; but that the sign of true progress is in its uninterruptedness and continual acceleration."*

There IS a Planet B

St. Francis of Assisi wrote, *"If you have men who will exclude any of God's creatures from the shelter of compassion, you will have men who deal likewise with their fellow men."*

Sir Thomas More said, *"Slaughtering our fellow creatures destroys our sense of compassion, which is the finest sentiment of which our human nature is capable."*

The Portals of Veganism

We come to Veganism through the portals of health, environment, ethics and spirituality.

While a few people come to Veganism for spiritual uplift, it is an important facet of Veganism that every Vegan appreciates eventually.

Compassion is undifferentiated, like a flower that shares its fragrance with all creatures equally. This is the ethical facet of Veganism and roughly one out of three[1] come to Veganism because they love animals.

Equally, about one out of three come to Veganism for their personal health, because they love themselves. Our modern environment is so full of toxins that eating animal foods has become increasingly hazardous for our health.

Animals store toxins in the fat cells of their bodies and it takes the average mammalian liver roughly 7 years to filter out and excrete half the toxins. How could our overworked liver possibly get rid of these toxins if we are consuming a fresh load of toxins in animal foods three times a day?

In contrast, plant foods[2] contain a factor of 10 to 100 less concentrations of toxins than comparable animal foods.

Finally, one out of three come to Veganism for environmental reasons, because they love our Mother Earth and they recognize that animal husbandry is the leading cause of the nature and climate crisis we face today.

After I embraced Veganism for environmental reasons in 2008, I began to appreciate the health, ethical and spiritual facets within a few months thereafter.

Today, I am an ethical, whole-foods, plant-based Vegan because I love myself, I love the animals and I love our Mother Earth. I'm also a Human Earth Animal Liberation (HEAL) advocate and activist.

I also believe that everyone is a Vegan at heart, since I have never met a single person who would deliberately hurt an innocent animal unnecessarily. Therefore, there are two kinds of people:
1. Conscious Vegans; and
2. Closeted Vegans

When we come out for Veganism on health, spiritual or ethical grounds, there is not the same sense of urgency as when we come out for Veganism on environmental grounds.

We might think, *"I'm feeling better today and therefore, I can consume animal products."* Or, *"what's the hurry if we have been exploiting animals for 10,000 years already?"*

But when I embraced Veganism for the environment, I felt a tremendous sense of urgency.

It was as if my house was on fire.

---

[1] The 2019 Global Survey on Veganism revealed these broad reasons on why people go Vegan. The link https://vomad.life/survey/ accessed on March 4, 2025.

[2] Please see the book, *Dioxins and Dioxin-like Compounds in the Food Supply: Strategies to Decrease Exposure,* published by the National Academies Press, 2004. Link https://nap.nationalacademies.org/catalog/10763/dioxins-and-dioxin-like-compounds-in-the-food-supply-strategies accessed on March 4, 2025.

5. House on Fire

"Set your life on fire.
Seek those who fan your flames."
– Rumi

When our house is on fire, we abandon our daily routines and do what it takes to put out the fire and save all that is truly important to us.

When our house is on fire, we don't worry about what's for breakfast or what we are using to cream our coffee. When our house is on fire, we don't throw lighted matches on the fire.

When our house is on fire, we tell everyone in the house that it is on fire. If our sister is in the next room unaware that the house is on fire, we don't say, "I won't disturb her. Let her find out for herself."

We tell her and together, we do what it takes to put out the fire.

My house did catch fire in 1993 and it taught me some important lessons on how our climate *heating* system on Planet A actually works.

The fire started in the garage of our home. It was a brand new home in suburban New Jersey, USA, and it was barely a month since we moved in. I had just bought a brand new car and parked it in the garage the previous night. The next morning, the car caught fire spontaneously and burned the house down to the point where we had to move out for eight months while the house was rebuilt.

There IS a Planet B

When the fire started, I was on the phone speaking to a colleague. Our two sons came running out of the room above the garage and said that there was smoke coming from the heating vents. I told them that the heater just came on and it would go away. I was too busy with my call to even investigate the cause of the smoke.

They came running out again moments later saying that there was black smoke coming out of the vents. At that point, the phone went dead and I knew that there was something really wrong. I ran downstairs and opened the door to the garage to discover the car spitting flames at me. There was quite a conflagration going on in the garage in front of the car.

At that point, my only thought was to call the Fire department for help and get our children to safety. I wasn't interested in my wallet or my passport or any of the other material things in the house, but just the precious lives that were in danger.

The Fire Marshall came to our house to investigate the cause of the fire and he determined that it started in the car. At that point, I called up the car dealer to ask him, *"What did you sell me?"*

His response shocked me. He said, *"Oh my God, it happened to you?!"*

Then he gave me the phone number of an engineer working for the car manufacturer and asked me to contact him.

I contacted the engineer and he informed me that he had been studying this spontaneous combustion problem for a year or so. Mine was the eighth car that caught fire, while headquarters was waiting for ten cars to catch fire before they issued a recall to fix the problem.

The problem had been fixed in all cars manufactured after May 1993 with a simple $1 plastic sleeve around a fuel line. My car had been manufactured in April 1993 and did not have that fix.

The spontaneous combustion occurred because the manufacturer had packed an eight-cylinder engine in a six-cylinder compartment and in the process, the designers had routed the fuel line too close to

the windshield washer heater. The windshield washer heater was always on and if the fuel line had a slow leak, fuel vapors accumulated around the heater and reached ignition point when the ambient temperature rose during the day. Cars with this fuel leak problem caught fire when parked in a closed space with no air flow, such as our garage.

As an engineer, I am well aware that no human design is perfect and flaws are bound to occur. Therefore, I was prepared to write the house fire off as my bad luck, but what happened next opened my eyes to how things actually work on Planet A.

The day after the fire, the car manufacturer sent a Professor of Physics from the University of Pennsylvania to examine the remains of our house fire. His stated purpose was to find the root cause of the fire.

I noticed that he was looking at everything in the garage except the car. I asked him, *"Professor, why are you looking at all these other things when I already spoke to Eric and he told me exactly how my fire happened?"*

The good professor got very agitated. I overheard him complaining to his team members, *"How can I do my job if people are giving out Eric's number to customers?"*

From that point onward, I couldn't reach Eric, the Engineer on the phone. He seemed to have just vanished. Then the Professor cooked up a theory that the gas heater that was in the corner of the garage caught fire and threw flames at the car to make it look like the car was the cause of the fire.

Our home insurance company, armed with the Fire Marshall's report, did not buy the Professor's fanciful theory and sued the car manufacturer for damages. The case dragged on for four years, at the end of which, it became a $4M lawsuit, even though our house cost far less to rebuild. After four years, the car manufacturer paid the insurance company $1.6M for a no-fault settlement.

There IS a Planet B

Meanwhile, two other cars also caught fire within a couple of months after our house fire and I received a recall notice from the car manufacturer to bring the car in for installing the $1 plastic sleeve around the fuel line to resolve the spontaneous combustion problem.

Just think about how much easier it would have been if everyone had told the truth right from the outset.

Of course, if I had known the car had that spontaneous combustion manufacturing defect, I would not have bought it. However, in an open source system on Planet B, as soon as the car manufacturer knew enough to install a plastic sleeve around the fuel line in May 1993, they would have recalled all the cars manufactured prior to May 1993 to retrofit them with the necessary fix, instead of making a cold, financial calculation about ten cars catching fire before issuing a recall.

6. Our Planetary Home on Fire

"Our house is still on fire. Your inaction
is fueling the flames by the hour."
– Greta Thunberg

Today, we have the same emergency situation in our planetary home. Our planet is truly on fire. I have been working on the nature and climate crisis since 2006 and every year, I see the fire alarms in our planetary home getting louder and louder.

After an unprecedented streak of 12 monthly temperature records[1] where every month was the hottest month ever recorded, in June 2024, Bill Nelson, the Administrator of NASA said, *"It is clear we are facing a climate crisis. Communities across America — like Arizona, California, Nevada — and communities across the globe are feeling first-hand extreme heat in unprecedented numbers."*

In December 2023, over 200 health journals[2] called on the UN, political leaders and health professionals to recognize that the climate and nature crises we face today are truly one indivisible crisis and must be tackled together to preserve health and avoid catastrophe and that this overall environmental crisis is now so severe as to be a global public health emergency.

There IS a Planet B

In 2009, a team of 26 scientists[3] under the aegis of the Stockholm Resilience Center unveiled the Planetary Boundaries Framework to explain how humans are impacting life support systems on the planet. In the framework, there are nine planetary boundaries that we must stay within if we wish to maintain life as we know it on the planet. If we transgress any of the boundaries, then the framework establishes a color-coded risk profile, ranging from yellow to orange to red in order to indicate the risk probability that the transgression might end life as we know it on the planet.

The nine planetary boundaries are:

1. Biospheric integrity: A key measure of humanity's global impact is the rate at which wildlife is dying out on this planet, both in terms of species extinction rates and in the reduction in the total populations of wild animals. Humans impact biospheric integrity through habitat loss, habitat fragmentation, introduction of invasive species, chemical pollution, depletion of freshwater sources and climate change.

Species extinction rates are currently estimated to be 1,000 to 10,000 times the background rate and wild animal populations are declining exponentially making biospheric integrity the worst of the planetary boundary transgressions.

2. Novel entities or Chemical pollution: Humans have introduced over 140,000 novel chemicals into the environment, mainly over the past century. We produce over 2 billion tons of these synthetic chemicals every year and in addition, the toxic waste products from our industrial processes total to about 220 billion tons of toxic chemicals[4] being poured into the environment every year.

This is a vast chemical experiment that is threatening the well-being of all life on earth. These chemicals are accumulating in the environment and working their way up the food chain in higher and higher concentrations, year after year. While trees and plants are very good filters and store the chemical pollution in their trunks and stalks, animal livers require seven years to

clean out half the toxins in our bodies. The problem gets compounded when we animals consume a fresh load of chemical pollution at every meal.

Next to habitat loss, chemical pollution is causing more destruction of wildlife than any other factor.

3. Nitrogen and Phosphorous loading: Humans have been impacting the nitrogen and phosphorous cycles of the planet through the over production and over use of synthetic fertilizers. Fertilizer runoff leads to dead zones in rivers and in the ocean as the over fertilized river and ocean beds lead to excess plant growth depleting the oxygen in the water and suffocating fish life.

At present, it is estimated that our nitrogen and phosphorous use is at least double the tolerable limits and this is also leading to the loss of biospheric integrity.

4. Climate change: Climate change is caused by an imbalance in the incoming solar energy and the outgoing thermal energy of the planet due to human additions of greenhouse gases and cooling gases in the atmosphere as well as surface albedo changes. The primary greenhouse gases that humans emit are carbon dioxide (CO_2), methane (CH_4) and nitrous oxide (N_2O), while the primary cooling gases are sulphur dioxide (SO_2), nitrogen oxides (NOx) and organic carbon.

In the absence of any other greenhouse gases, it is estimated that CO_2 concentrations in the atmosphere must be below 350 parts per million (ppm) for climate change to be in the green zone. At present the CO_2 concentration in the atmosphere is at 424 ppm, transgressing this limit.

Human CO_2 additions to the atmosphere have caused[5] 2.06 W/ m^2 of global warming in terms of Effective Radiative Forcing (ERF). That is, the 144 ppm of added CO_2 in the atmosphere from 1750 levels (280 ppm) is like adding a 2.06 Watt heater on every square meter of the earth's surface. However, other

There IS a Planet B

greenhouse gases such as methane, nitrous oxide, etc., together contribute even more, 2.2 W/m^2 of ERF to global warming. Cooling gases such as sulphur dioxide together reduce this global warming by 1.66 W/m^2 of ERF so that the net human caused global warming ERF is 2.6 W/m^2.

Recently, James Hansen and his colleagues have deduced from the pattern of rise in ocean surface temperatures during 2023-24 that the Charney sensitivity[6], which is the expected rise in global surface temperatures due to the doubling of CO_2 concentration in the atmosphere, is 4.5°C instead of the 3°C used by the UN IPCC. In that case, the ERF due to CO_2 increases from 2.06 W/m^2 to 3.09 W/m^2 and correspondingly, the ERF from cooling gases increases from 1.66 W/m^2 to 2.69 W/m^2.

In terms of ERF, the green zone for climate change is transgressed at 1 W/m^2. Climate change impacts biospheric integrity mainly through requiring species migrations at rates beyond what is feasible to maintain the integrity of the ecosystem.

5. Land system change: This planetary boundary is impacted by the transformation of native ecosystems into grazing land, timber land, crop land or built land. At present, only 9% of the ice-free land area of the planet is covered by native ecosystems making land system change a major factor in the loss of biospheric integrity.

6. Freshwater change: Changes in fresh water flows and the availability or lack thereof of fresh water at root systems impact the well being of life. Humans have been modifying fresh water flows through straightening out and damming rivers and the destruction of forests. Humans have also been impacting the availability of fresh water at root systems through irrigation.

7. Ocean acidification: As CO_2 levels in the atmosphere increase, more of that CO_2 dissolves in the ocean to become H_2CO_3 or carbonic acid, thereby increasing the acidity of the ocean. As the water becomes more acidic, sea animals that grow

shells find it more difficult to survive. At present, ocean acidification lies at the margin of the safe operating space, but the trend is worsening as the CO_2 levels in the atmosphere increase year on year.

8. Atmospheric aerosol loading: Dust and other microscopic particles in the atmosphere impact biospheric integrity, both by masking sunlight and causing breathing difficulties for animals. In addition, asymmetries in aerosol loading in the northern and southern hemispheres can affect multiple monsoon systems. Humans have been changing the aerosol loading in the earth's atmosphere mainly through the emissions of SO_2 which leads to the formation of sulphate aerosols (H_2SO_4) and the cloud interactions that result from these aerosols in the atmosphere.

9. Stratospheric ozone depletion: The stratospheric ozone layer, which is the prevalence of ozone in the layer of the atmosphere 20 km to 50 km from the earth's surface, protects life on earth from the harmful UltraViolet (UV) rays of the sun. The formation of this ozone layer in the upper atmosphere was necessary for complex life to thrive on planet Earth. The Montreal protocol was signed by the nations of the world in 1987 when it was discovered that a class of chemicals called ChloroFluoroCarbons (CFCs), used mainly in refrigeration, were causing the stratospheric ozone layer to thin out, especially over the poles.

Since then, according to the World Meteorological Organization (WMO), the ozone holes over the poles have been closing at the rate of 1-3% per decade. The WMO estimates that if current policies remain in place, the ozone layer is expected to recover to 1980 values by 2066 over the Antarctic, by 2045 over the Arctic and by 2040 over the rest of the world.

In 2009, the scientists from the Stockholm Resilience Center (SRC) assessed that humans were in the red zone on two planetary boundary transgressions: biospheric integrity and nitrogen and phosphorous loading and in the orange zone on climate change.

There IS a Planet B

In 2015, the SRC team issued an update[7] that humans were still in the red zone on two planetary boundary transgressions – biospheric integrity and nitrogen and phosphorous loading, but in the orange zone on two planetary boundary transgressions: climate change and land system change.

As of 2023, the nature and climate crisis had become progressively worse and now humans were in the red zone on four planetary boundary transgressions[8] – biospheric integrity, chemical pollution, nitrogen and phosphorous loading and climate change and in the orange zone on two planetary boundary transgressions – land system change and freshwater change.

This was like a 3-alarm planetary fire in 2009, which became a 4-alarm fire in 2015 and as of 2023, it was a raging 6-alarm fire. How we respond to this planetary fire determines what kind of future we can expect on planet earth. On Planet A, we have been responding to it largely in a self-interested manner, pretending to put out the fire while actually enriching ourselves, whereas on Planet B, we are responding to it accurately.

---

[1] *NASA Analysis Confirms a Year of Monthly Temperature Records* was issued by NASA in May 2024. The link https://www.nasa.gov/earth/nasa-analysis-confirms-a-year-of-monthly-temperature-records/ accessed on March 4, 2025.

[2] Abbasi, et al., *Time to Treat the Climate and Nature Crisis as one Indivisible Global Health Emergency,* Nature 2023. Link https://www.nature.com/articles/s41533-023-00358-3 accessed on March 4, 2025.

[3] Rockstrom, et al., *A Safe Operating Space for Humanity,* Nature, Sep 2009. Link https://www.nature.com/articles/461472a accessed on March 4, 2025.

[4] Naidu, et al., *Chemical Pollution: A Growing Peril and Potential Catastrophic Risk to Humanity,* Environment International, Vol 156, November 2021. Link https://www.sciencedirect.com/science/article/pii/S0160412021002415 accessed on March 4, 2025.

[5] Forster P., et al. *The Earth's Energy Budget, Climate Feedbacks, and Climate Sensitivity*. In: Climate Change 2021: The Physical Science Basis Contribution of Working Group I to the Sixth Assessment Report of the Intergovernmental Panel on Climate Change [Internet]. Cambridge University Press, Cambridge, United Kingdom and New York, USA; 2021. Link https://www.ipcc.ch/report/ar6/wg1/chapter/chapter-7/ accessed March 6, 2025.

[6] Hansen JE, et al. *Global Warming Has Accelerated: Are the United Nations and the Public Well-Informed?* Environment: Science and Policy for Sustainable Development. 2025 Jan 2;67(1):6–44. Link https://www.tandfonline.com/doi/full/10.1080/00139157.2025.2434494#d1e751 accessed March 6, 2025.

[7] Steffen, W., et al., *Planetary Boundaries: Guiding Human Development on a Changing Planet,"* Science, vol. 347, no. 6223, Jun 2015. Link https://www.science.org/doi/10.1126/science.1259855 accessed March 6, 2025.

[8] Richardson, K., et al., *Earth Beyond Six of Nine Planetary Boundaries,* Science Advances, vol. 9, no. 37, Sep 2023. Link https://www.science.org/doi/10.1126/sciadv.adh2458 accessed March 6, 2025.

7. A Critical Look at the UN IPCC's Emissions Accounting

*"What do you get when you mix science
with politics? Pure politics."*
- John Barry

We have not been responding effectively to the planetary fire until now. On Planet A:

**We've run out of solutions
So don't tell us that
We have been misled by the UN IPCC
Because the reality is
Rapidly phasing down fossil fuel use should be a top priority
And we don't believe anyone who says
Animal husbandry is the leading cause of climate change**

(On Planet B, we upend the narrative by reading this from the bottom up)

In the mainstream scientific community, most of the six planetary boundary transgressions get very little attention, evidently because animal husbandry is well established to be the leading cause of that transgression or ending animal husbandry is the leading solution to that transgression. This is not because scientists are not interested in studying transgressions involving animal husbandry, but because

There IS a Planet B

there is very little funding in the climate *heating* system on Planet A for exploring any activity that would curtail animal husbandry.

The only transgression that gets the most attention is climate change because it is being framed as a fossil fuel issue, for which there is a "green growth" story. The "green growth" story goes that if we replace the fossil fuel infrastructure with electric cars, solar panels and other renewable energy infrastructure and thereby grow the human economy, we can solve climate change. This story was pitched relentlessly over the past three decades by mainstream climate spokespeople so that the richest man in the world today is someone who makes and sells electric cars.

This "green growth" story is not only false, but it is extremely dangerous. Animal husbandry is also the leading cause of climate change and it appears to have been only through creative accounting on emissions that the United Nations (UN) Intergovernmental Panel on Climate Change (IPCC) has been framing climate change as largely a fossil fuel problem.

The green growth story is dangerous because if we stop fossil fuel use now as climate spokespeople at the UN are advocating, it will **increase the global warming on the planet by 50-87% within a few weeks**, potentially triggering a number of positive feedback loops in the climate system and possibly runaway climate change. This is suicidal.

What is the UN IPCC?

The UN IPCC is an organization of governments that are members of the United Nations or the World Meteorological Organization (WMO). Created in 1988 by the WMO and the UN Environmental Program (UNEP), the purpose of the UN IPCC is to provide governments at all levels with scientific information that they can use to develop climate policies.

Thousands of scientists contribute to the work of the UN IPCC and it has a transparent and open process by which the work of the scientists is reviewed by experts and governments around the

43

world. Therefore, the corruption of the UN IPCC's reports is quite blatant as there is recorded evidence of political operatives significantly altering[1] language and conventions in the content of the scientific reports.

The Three Axioms

With regard to emissions accounting, there are three axioms:

1. **CO2 is CO2.** CO_2 is a well-mixed gas in the atmosphere and a CO_2 molecule behaves the same way no matter what its emission source.

2. **Photosynthesis occurs through the grace of Nature.** There is no such thing as anthropogenic photosynthesis. It is arrogant and even incorrect from an engineering perspective for humans to take credit for photosynthesis and use it to offset certain kinds of emissions, if we are serious about addressing the nature and climate crisis.

3. **Emissions accounting should include ALL emissions.** We cannot draw conclusions considering just a few emissions species, while ignoring several other equally relevant emissions species.

The UN IPCC violates all of these axioms in its sector emissions accounting.

Sector/ Species	Fossil Fuels	Industry	Agriculture	Forestry	Waste	Other	Total (1750-2020)
CO_2	1700.1	43.4	646.8	201.0	0	73.9	2665.2
Methane	259.1	0.8	374.8	0.6	102.5	0.2	737.8
N_2O	16.3	16.2	147.5	1.4	6.7	19.2	207.5
Halocarbons	0	479.7	0	0	0	0	479.7
Total CO_2e	1975.6	540.1	1167.0	203.0	109.3	93.3	4090.3
Percentage	48%	13%	29%	5%	3%	2%	100%

Table 1: The UN IPCC's GWP100 sector emissions accounting utilizes just 4 emissions species.

There IS a Planet B

Four Ways to Politicize Emissions Accounting

There are four main ways in which the UN IPCC is politicizing the accounting on greenhouse gas emissions to make it seem like fossil fuel burning is the leading cause of climate change:

1. **The UN IPCC undercounts deforestation emissions by a factor of 3.** The UN IPCC undercounts deforestation and land use change CO_2 emissions by using *net* accounting for land use change emissions and *gross* accounting for fossil fuel emissions even though CO_2 is a well-mixed gas in the atmosphere and a CO_2 molecule emitted from deforestation behaves exactly the same as a CO_2 molecule emitted from fossil fuel combustion. This violates Axiom #1.

   With net accounting, emissions from deforestation in one area is offset by regrowth on managed or abandoned land in another area. While using net accounting, the UN IPCC is allowing humans to take credit for photosynthesis, which violates Axiom #2.

   When we use consistent[2] *gross* accounting for deforestation and land use change CO_2 emissions, the annual value increases from 1.6 GtC to 3.4 GtC and the cumulative emissions from 1750 to 2020 increases from 200 GtC to 553 GtC, by nearly a factor of 3. In contrast, fossil fuel combustion has caused 464 GtC of cumulative emissions from 1750 to 2020.

   When the UN IPCC undercounts deforestation and land use change emissions by a factor of 3, it underestimates the impact of animal agriculture and overestimates the impact of fossil fuel combustion.

2. **The UN IPCC undervalues methane emissions by a factor of 3.** The UN IPCC uses a 100 year averaging of the impact of methane on global warming in its sector accounting even though climate change is imminent and we don't have the

45

luxury of waiting 100 years to solve the nature and climate crisis. Methane has a half-life of less than 10 years in the atmosphere as it reacts with hydroxyl radicals in the atmosphere to become CO_2. The 100 year Global Warming Potential (GWP100) of methane is 28, implying that every Gt of methane traps as much heat as 28 Gt of CO_2 in the atmosphere when we consider its impact over 100 years.

The UN IPCC has also assessed that methane has caused 1.2 W/m^2 of anthropogenic global warming Effective Radiative Forcing (ERF) on the planet while CO_2 has caused 2.06 W/m^2, measured cumulatively from 1750 to 2020. Using consistent accounting of land use change CO_2 emissions, we can calculate that each Giga ton of CO_2 has caused 2.06/4216 = 0.000489 W/m^2 of ERF, while each Giga ton of methane has caused 1.2/26.4 = 0.0455 W/m^2 of ERF or the GWP-ERF of methane is 0.0455/0.000489 = 93, more than 3 times the GWP100 value of 28 used by the UN IPCC in its sector accounting[3].

When two sets of data issued by the UN IPCC are inconsistent, it is the set based on its *arbitrary choice* of 100 years for averaging the impact of methane that needs to be discarded and ignored. Hence, by using the GWP100 metric for their sector emissions comparisons, the UN IPCC has been undervaluing the impact of methane emissions by a factor of 93/28 = 3.32.

When the UN IPCC undervalues methane emissions by a factor of 3, it undervalues the impact of animal agriculture and overvalues the impact of fossil fuel combustion.

3. **The UN IPCC ignores cooling effects.** The UN IPCC ignores the cooling effects that occur mainly due to fossil fuel combustion in its sector emissions accounting. The cooling gases co-emitted with fossil fuel combustion result in 1.3 W/m^2 of cooling ERF from fossil fuel sources alone. These

There IS a Planet B

cooling gases, SO_2, NOx and Organic Carbon, are very short-lived with a half-life of a few weeks at most. Therefore, if humanity rapidly phases down fossil fuel use without addressing animal husbandry as most prominent climate spokespeople are recommending, this would be the exact opposite of what we need to do to address climate change since it would cause anthropogenic global warming **ERF to rapidly increase from 2.6 W/m^2 to 3.9 W/m^2, a 50% increase in a matter of a few weeks.**

When the UN IPCC ignores cooling effects, it violates Axiom #3 and validates government policies that are suicidal for all life on earth. When the UN IPCC ignores cooling effects, it is behaving like a little child who is hiding a half-eaten popsicle in one hand behind his back, claiming that he had nothing to do with the disappearance of the popsicle.

4. **The UN IPCC ignores the Carbon Opportunity Cost (COC) of the land used for animal husbandry.** The UN IPCC ignores the vast majority of anthropogenic CO_2 emissions associated primarily with animal husbandry activities by claiming that they are part of the "natural" cycle. In the Sixth Assessment Report of the UN IPCC (AR6), animal respiration, pasture maintenance fires, bottom trawling of the ocean and other such "natural" activities were shown to be responsible for **48.6** GtC of emissions[4], more than five times as much as the **9.4** GtC emitted through fossil fuel combustion.

This presumes that the artificial impregnation of billions of farmed animals is a natural process for which humans bear no responsibility.

This presumes that bottom trawling of the ocean with 100 mile long nets is a natural process for which humans bear no responsibility. This presumes that chopping down and burning vegetation that the farmed animals did not eat in order to maintain grazing lands and prevent forests from regenerating is a natural process for which humans bear no responsibility.

On the other hand, if we humans accept responsibility for bringing all these farmed animals into this world, then we humans also have to take responsibility for all the CO_2 emissions of our animal husbandry activity.

The main rationale for not counting emissions from animal respiration and pasture maintenance fires is that they were accounted for when the land was deforested. However, when the land is repeatedly attempting to regenerate year after year and the farmed animals consume that vegetation or it gets burnt up for pasture maintenance, the overall CO_2 emissions from that land would far exceed the original carbon stored on that land.

The UN IPCC also ignores historic deforestation and land use change CO_2 emissions for the 10,000 years of agriculture in the pre-industrial era prior to 1750 on the grounds that climate change did not happen prior to 1750.

This is like claiming that the hamburgers we have been eating for 50 years had nothing to do with the heart disease we have contracted because we only felt chest pains after we also started drinking milkshakes two years ago.

This also violates Axiom #2, specifically as humans would be taking credit for growing peat moss in the Arctic to compensate for our deforestation emissions. Therefore, whether climate change happened or not, CO_2 emissions is CO_2 emissions (Axiom #1) and must be counted.

Since the "natural" CO_2 emissions and historic CO_2 emissions were primarily caused by animal husbandry, this once again shows that the UN IPCC has been politicizing emissions accounting in order to reduce the impact of that sector.

One way to account for all these emissions is to include the **Carbon Opportunity Cost** (COC) of the land used for animal

There IS a Planet B

husbandry which is the cost we incur for choosing to continue this unnecessary activity, when we could alternately be regenerating ecosystems and storing carbon on that same land.

Sector/ Species	Fossil Fuels	Industry	Animal Agriculture	Other Agriculture	Forestry	Waste	Other	Total ERF (W/m^2) 1750-2020
CO_2	0.830	0.021	0.786	0.108	0.277	0.000	0.036	2.058
Methane	0.393	0.001	0.561	0.066	0.001	0.172	0.000	1.195
N_2O	0.019	0.019	0.102	0.070	0.002	0.008	0.022	0.241
Halocarbons	0.0	0.211	0.0	0.0	0.0	0.0	0.0	0.211
NMVOC+CO	0.283	0.013	0.091	0.014	0.005	0.003	0.031	0.440
Black C	0.082	0.0	0.016	0.003	0.001	0.001	0.005	0.107
SO_2	-0.775	-0.131	-0.012	-0.001	-0.001	-0.001	-0.018	-0.939
NOx	-0.247	-0.003	-0.010	-0.003	0.000	-0.003	0.000	-0.267
NH_3	-0.004	0.0	-0.023	-0.005	0.000	0.000	-0.001	-0.034
Organic C	-0.105	0.0	-0.057	-0.006	-0.004	-0.003	-0.034	-0.209
Albedo	0.0	0.0	-0.112	-0.040	-0.016	0.000	-0.032	-0.200
Total ERF	0.475	0.131	1.341	0.206	0.264	0.177	0.009	2.603
Percentage	18%	5%	52%	8%	10%	7%	0%	100%

Table 2: Consistent ERF sector emissions accounting utilizing all 11 emissions species (Charney Sensitivity for CO_2 doubling = 3°C) – From Wedderburn-Bisshop (2025).

The UN IPCC is validating suicidal government policies

This is the Cow in the Room at the UN IPCC, which is using inconsistent and biased accounting conventions to artificially suppress the impact of animal husbandry and amplify the impact of fossil fuel combustion in order to promote a "green growth" story for renewable energy.

Mainstream climate spokespeople are now promoting this "green growth" story and recommending the rapid phaseout of fossil

fuels[5]. This is dangerous because the world can expect a **50% increase in global warming ERF in the short term** if such rapid phaseout recommendations are actually implemented without addressing the Cow in the Room as most government policies are currently configured.

This is suicidal.

If and when Hansen et al.'s calculation[6] that the cooling impact of SO_2 has been underestimated by 1 W/m^2 in IPCC AR6 is validated, then the cooling gases co-emitted with fossil fuel combustion would result in 2.3 W/m^2 of cooling ERF from fossil fuel sources alone. Then, phasing down fossil fuel use rapidly would cause anthropogenic global warming ERF to rapidly increase from 2.6 W/m^2 to 4.9 W/m^2, **an 87% increase in a matter of a few weeks**.

Remarkably, with Hansen et al.'s update, the heating effects of fossil fuels and industry are almost exactly cancelled out by their cooling effects so that these two sectors have collectively caused just 2% of the global warming ERF from 1750 to 2020.

There IS a Planet B

Sector/ Species	Fossil Fuels	Industry	Animal Agriculture	Other Agriculture	Forestry	Waste	Other	Total ERF (W/m^2) 1750-2020
CO_2	1.245	0.032	1.179	0.162	0.415	0.000	0.054	3.087
Methane	0.393	0.001	0.561	0.066	0.001	0.172	0.000	1.195
N_2O	0.019	0.019	0.102	0.070	0.002	0.008	0.022	0.241
Halocarbons	0.0	0.211	0.0	0.0	0.0	0.0	0.0	0.211
NMVOC+CO	0.283	0.013	0.091	0.014	0.005	0.003	0.031	0.440
Black C	0.082	0.0	0.016	0.003	0.001	0.001	0.005	0.107
SO_2	-1.624	-0.275	-0.026	-0.002	-0.001	-0.002	-0.038	-1.968
NOx	-0.247	-0.003	-0.010	-0.003	0.000	-0.003	0.000	-0.267
NH_3	-0.004	0.0	-0.023	-0.005	0.000	0.000	-0.001	-0.034
Organic C	-0.105	0.0	-0.057	-0.006	-0.004	-0.003	-0.034	-0.209
Albedo	0.0	0.0	-0.112	-0.040	-0.016	0.000	-0.032	-0.200
Total ERF	0.041	-0.003	1.721	0.259	0.402	0.176	0.007	2.603
Percentage	2%	0%	66%	10%	15%	7%	0%	100%

Table 3: Consistent ERF sector emissions accounting utilizing all 11 emissions species (Charney Sensitivity for CO_2 doubling = 4.5°C from Hansen et al (2025)).

Even a 50% increase in global warming ERF, let alone an 87% increase, would result in the global surface temperature of the Earth crossing the 2°C threshold within a few years. According to Hansen et al., due to the inertia of the ocean, one-third of the impact of this climate forcing will be felt within 5 years, the next one-third over the next 100 years and the final one-third over millennia.

The Seven Climate Tipping Points at 1.5°C

There are at least seven climate tipping points[7] — defined to be conditions beyond which changes to a part of the climate system become self-perpetuating, — that are **likely to be triggered** as the global surface temperature increases beyond the 1.5°C threshold:

1. **Boreal Permafrost Abrupt Thaw:** Permanently frozen soils in the northern boreal regions lock in over a trillion tons of

carbon that can be released as CO_2 and methane upon thawing. The mean estimate for the global surface temperature increase that can lead to an an abrupt thaw of the boreal permafrost is 1.5°C.

2. **Low Latitude Coral Reefs Die-off:** When water temperatures exceed a certain threshold, tropical and sub-tropical coral reefs expel their algae and bleach resulting in the death of one of the most biodiverse ecosystems on our planet. The mean estimate for the global surface temperature increase that can lead to widespread bleaching of low latitude coral reefs is 1.5°C.

3. **Labrador Sea Sub-polar Gyre Collapse:** Ocean current circulation in the North Atlantic abruptly collapses in some models at a global surface temperature increase of above 1.8°C. This could result in a potential 2-3°C decrease in average temperatures in the North Atlantic nations of the world.

4. **West Antarctic Ice Sheet Collapse:** The West Antarctic Ice Sheet (WAIS) is perched on land but large parts of it are submerged under the warming ocean. It is estimated that many glaciers within WAIS will experience a form of self-sustaining retreat at global surface temperature increases above 1.5°C.

5. **Barents Sea Winter Ice Collapse:** The Barents Sea Ice in the Arctic has been shown to collapse even during winter in models at global surface temperature increases above 1.5°C. These models also project occasional Septembers with no Arctic Summer ice at similar temperature increases.

6. **Greenland Ice Sheet Collapse:** Models as well as paleoclimate data show that the Greenland ice sheet could reach a tipping point and melt rapidly due to a positive feedback loop that forms as the height of the ice sheet decreases, the surface ice encounters warmer air and melts

There IS a Planet B

faster. This feedback loop is triggered in models at global surface temperature increases above 1.5°C.

7. **Amazon Forest Die-Back:** Taking into account both biological and physical feedback systems, the Amazon rainforest is calculated to die-back and turn into a savannah at global surface temperature increases of 1.5-2°C.

All these seven climate tipping points are poised to trigger at global surface temperature increases of 1.5°C above pre-industrial levels and therefore, it makes no sense to pursue a rapid phaseout of fossil fuel use instead of what a competent systems engineering team would recommend: a rapid phaseout of animal husbandry and a more nuanced phaseout of fossil fuel use.

As even the UN IPCC would agree, there is nothing that does not improve when we shut down the folly of animal husbandry.

Open Letter to the UN IPCC

Another problem with the UN IPCC politicizing greenhouse gas emissions accounting is that it makes not only all government policies, but also derivative calculations as well as the research work done by universities and NGOs based on this accounting largely incorrect. Ecological overshoot day calculations and carbon footprint and carbon market pricing calculators that use the UN IPCC data are also largely incorrect.

A number of us have now signed an open letter[8] to the chairman of the UN IPCC, Dr. Jim Skea, calling on him to correct these errors and thereby reclaim the scientific integrity of the UN IPCC from political interference. This would help preserve the reputation of the thousands of hard working climate scientists who contribute to the UN IPCC and other researchers who depend on its data.

At the moment, the UN IPCC is acting like the Professor from the University of Pennsylvania who attempted to frame the heater in the corner of my garage for our house fire instead of the spontaneously combusting car manufactured by his employer.

While the Professor's dissembling was conducted after my house fire was put out, the UN IPCC's dissembling is causing us to misdirect our resources and efforts while our planetary home is still on fire, with ongoing tragic consequences for all life on Earth.

On Planet B, we see through this dissembling and respond to the situation accurately.

---

[1] The main altering of language in the UN IPCC reports happens with reference to plant-based diets, as the UN IPCC is the vehicle through which the "green growth" story is being promoted. Here's an article explaining how this has been happening for years: Link https://qz.com/ipcc-report-on-climate-change-meat-industry-1850261179 accessed on March 6, 2025.

[2] Gerard Wedderburn-Bisshop, *Deforestation: A Call for Consistent Carbon Accounting*, Environmental Research Letters, vol. 19, no. 11, Oct 2024. This work uses the data found in Houghton, R.A., and Castanho, A., *Annual Emissions of Carbon from Land Use, Land-Use Change and Forestry from 1850 to 2020*, Earth Syst. Sci. Data, Vol 15, 2023. Links https://iopscience.iop.org/article/10.1088/1748-9326/ad7d21 and https://essd.copernicus.org/articles/15/2025/2023/ accessed on March 7, 2025.

[3] Gerard Wedderburn-Bisshop, *Increased Transparency in Accounting Conventions Could Benefit Climate Policy*, Environmental Research Letters, vol. 20, no. 2, Mar 2025. Link https://iopscience.iop.org/article/10.1088/1748-9326/adb7f2 accessed on March 7, 2025.

[4] Canadell, P., et a., *Global Carbon and Other Biogeochemical Cycles and Feedbacks* in Climate Change 2021 – The Physical Science Basis: Working Group I Contribution to the Sixth Assessment Report of the Intergovernmental Panel on Climate Change, Cambridge: Cambridge University Press; 2021. Link https://www.ipcc.ch/report/ar6/wg1/downloads/figures/IPCC_AR6_WGI_Figure_5_12.png accessed on March 7, 2025.

There IS a Planet B

---

[5] The State of the Climate Report is an annual report published by Oxford Academic BioSciences and authored by William Ripple, et al. in 2024. Link https://academic.oup.com/bioscience/article/74/12/812/7808595?login=false accessed on March 7, 2025.

[6] Hansen JE, et al. *Global Warming Has Accelerated: Are the United Nations and the Public Well-Informed?* Environment: Science and Policy for Sustainable Development. 2025 Jan 2;67(1):6–44. Link https://www.tandfonline.com/doi/full/10.1080/00139157.2025.2434494#d1e751 accessed March 6, 2025.

[7] Armstrong-McKay, D. I., et al., *Exceeding 1.5°C Global Warming Could Trigger Multiple Tipping Points*, Science, vol. 377, no. 6611, Sep. 2022. I have also included the Amazon Rainforest Die-Back as the 7th potential climate tipping event based on the Grantham Lecture by Prof. Johan Rockstrom at the Divecha Center for Climate Change, Indian Institute of Sciences, Feb 2025. Link https://www.science.org/doi/10.1126/science.abn7950 accessed on March 7, 2025.

[8] Please see https://climatehealers.org/un-ipcc-open-letter/, accessed on March 7, 2025.

8. The Worst Planetary Boundary Transgression

"If we save the living environment, the biodiversity that we have left,
then we will also automatically save the physical environment.
If we only save the physical environment, then
we will ultimately lose both."
– E. O. Wilson

Of the six planetary boundary transgressions, the worst is biospheric integrity, specifically, the rate of loss of wild animals and other species on the planet. The World Wildlife Fund (WWF) Living Planet Report (LPR) has been monitoring the loss of wild animals through a statistical survey of thousands of representative species who live on land, in the water and in the air. In 2014, it reported[1] that between 1970 and 2010, we wiped out **52%** of all wild animals by total weight.

It takes four years for the WWF to compile the data and therefore, it reported these 2010 statistics in 2014. When that report came out, I extrapolated this decline in the total weight of wild animals assuming that we were killing them proportional to the size of the human economy. In 2014, the global Gross Domestic Product (GDP) was growing at 3% per year[2] and my crude extrapolation showed that we were on track to a world with virtually 100% loss of wild animals by 2026.

There IS a Planet B

I was shocked when I made this calculation. Later, I reported on this calculation at a public event where I pledged to ensure that there will be more wild animals in 2026 than in 2010. That public pledge was greeted with a standing ovation.

At the end of my presentation, several of the audience members approached me to ask how they could help me keep this pledge. I told them that the most important thing they could do was to go vegan and to help their family and friends to go vegan.

They asked if there was anything else they could do to help me keep this pledge. It was clear that going vegan was not on their radar screen at all.

At that point, I foolishly backed down. I thought that my model was quite crude and perhaps I should wait for the next WWF Living Planet Report to be issued, before taking any further steps. The WWF Living Planet Reports are issued once every two years and the next one was going to be published in 2016 and would report on 2012 statistics.

I deeply regret this decision today. I see my inaction as no different from the car manufacturer waiting for ten cars to catch fire before issuing a recall to fix the spontaneous combustion defect in the car.

It also reeks of **speciesism**, the belief that one species, specifically humans, is more important than others. Imagine if I had done that during the Nazi era regarding the rate at which the Jewish population was being murdered or if I do that today regarding the rate at which the Palestinian population is being murdered!

Besides, the next report confirmed the usefulness of my model. The WWF LPR 2016[3] reported that between 1970 and 2012, wild animal populations declined by **58%** — a 6% rise in just two years. At that point, I knew that my model was reasonably on target and this was the genesis of the Vegan World 2026[4] project at Climate Healers.

This project is based on the real possibility that unless the world goes largely vegan and causes an increase in the population of wild

animals year on year by the end of 2026, the ecological noose around humanity's neck can never be loosened.

Since 2016, all the WWF Living Planet reports have shown depressing reductions in wild animal populations and the latest one issued in 2024[5] showed that between 1970 and 2020, wild animal populations declined by **73%**.

We also know that between 10,000 years ago and 1970, wild animal populations had declined **60%** by total weight[6]. If we assume business as usual reduction in wild animal populations between 2020 and 2024, then their populations are likely down **94%** from pristine levels by the end of 2024.

With respect to this horrific loss of biodiversity, the late Prof E. O. Wilson of Harvard University said:

*"The worst thing that will probably happen – in fact, is already underway – is not energy depletion, economic collapse, conventional war or the expansion of totalitarian governments. As terrible as these catastrophes would be for us, they can be repaired in a few generations. The one process going on now that will take millions of years to correct is the loss of genetic and species diversity by the destruction of natural habitats. **This is the folly our descendants are least likely to forgive us.**"*

---

[1] The World Wildlife Fund Living Planet Report documents the state of the planet —including biodiversity, ecosystems, and demand on natural resources—and what this means for humans and wildlife. The 2014 report can be accessed here: https://www.worldwildlife.org/pages/living-planet-report-2014, accessed on March 7, 2025.

[2] Here is the World Bank data on GDP growth over the years: https://data.worldbank.org/indicator/NY.GDP.MKTP.KD.ZG, accessed on March 7, 2025.

There IS a Planet B

---

[3] https://www.worldwildlife.org/pages/living-planet-report-2016, accessed on March 7, 2025.

[4] https://climatehealers.org/veganworld, accessed on March 7, 2025.

[5] https://www.worldwildlife.org/publications/2024-living-planet-report, accessed on March 7, 2025.

[6] From Barnosky, A., *Megafauna Biomass Tradeoff as a Driver of Quaternary and Future Extinctions,* Proceedings of the National Academy of Sciences, Aug 2008. Link https://www.pnas.org/content/105/Supplement_1/11543.full accessed on March 7, 2025.

9. Nature the Perfect System Design

"Each species is a masterpiece, a creation assembled with extreme care and genius."
– E. O. Wilson

From a systems perspective, I consider Nature to be the perfect system design. Every species belongs exactly as is and contributes to the well being of all life in an ecosystem, whether it knows it or not.

The waste from one species is the food for another. The elephant may seem like she is being destructive when she is breaking branches off trees, eating the leaves, throwing the branch away; then breaking another branch, eating the leaves and throwing the branch away.

When we look closely, we notice that wherever the elephant broke branches off trees, that's where the sunlight streams down to nourish the underbrush. If the elephant hadn't broken branches off trees, the forest canopy would have been so thick that the underbrush would have died from lack of sunlight. The branches that she threw away also become the home for numerous critters on the forest floor.

There IS a Planet B

At some point, the elephant stops breaking branches off trees and moves on to do something else. Wherever the elephant tramples on bushes as she moves on, that is where new pathways are formed in the forest.

Wherever the elephant drops huge mounds of dung, that is where new Jackfruit trees are born. The Jackfruit is perfectly designed for the elephant who can insert a whole, ripe fruit in her mouth, crunch through the thick skin and enjoy the fruit while swallowing the seeds.

The elephant then walks tens of miles before depositing the dung. The Jackfruit seeds appear unscathed in the dung along with the manure to nourish the young Jackfruit tree, thereby spreading these majestic trees throughout the Western Ghat forests.

Even though, at first glance, the elephant seemed to be destructive to the forest, when we take a closer look, we find that she belongs exactly as is.

Is there a similar redeeming story for human beings?

Even though we seem to be so destructive to the planet, are we somehow serving a larger purpose, a purpose that we are perhaps unconsciously fulfilling in the perfect system design that is Nature?

As part of that larger purpose, are we now being called to move on from our current obsession with resource extraction and wealth accumulation to do something else, just as the elephant was signaled to move on from breaking branches off trees?

When we take a closer look from a systems perspective, the answer to these three questions is indeed a resounding yes, yes, and yes!

In Nature, every ecosystem and every species organizes through a primary code that Dr. Shelley Ostroff calls, **The Vitality Code**[1]: *"All parts of a living system receive precisely what they need in order to manifest their unique potential in mutual nourishment with the interconnected whole. All living beings receive what they need, to give all they can, to become all they are."*

---

[1] Offered by Dr. Shelley Ostroff, the Vitality Code refers to Nature's primary code and principle of self-organizing and self-regulating thriving living systems. Link https://www.codes.earth/vitality-code accessed on March 7, 2025.

10. How Humans Belong in Nature

"We are biologically, cognitively, physically and
spiritually wired to love… and to belong."
– Brené Brown

The name, "Homo Sapiens," meaning "wise hominid" in Latin, was meant to highlight the unique intellectual abilities of human beings. Possessing a pre-frontal cortex, opposable thumbs and the unique superpower to start fires whenever we wish, we are perfectly positioned to make an extraordinary contribution to the well-being of all life, once we respond accurately to the six-alarm fire signal from Nature.

With our superpower of the control of fire, humans have been heating up the earth, especially over the past 10,000 years. We did that at first by cutting down trees and burning them to clear land for agriculture and for grazing farmed animals. Then, about 250 years ago, we discovered fossil fuels, dug them up and burnt them to fuel our industrial revolution and heated up the planet even more.

Now we have discovered that we have created a mess with too much pollution and we have overheated the planet. As soon as we discovered that we are actually capable of overheating the planet

There IS a Planet B

and disrupting the various life-support systems of the planet, we collectively became responsible for stabilizing the climate of the planet and maintaining all the life-support systems within their safe zones.

As Dr. Shelley Ostroff observed[1], *"The only legitimate purpose of governance is to protect and cultivate the health and vitality of the planet and all its inhabitants for generations to come."*

Whether we like it or not, we have become **Vitally Engaged Guardians of Animals and Nature** (VEGAN) and a life supporting, climate regulating "thermostat" species of the planet. We cannot now simply throw up our hands and expect the tiger or the monkey to take on these responsibilities.

Mother Earth has singled us out and tagged us. We are it!

The sooner we take this ecological responsibility seriously and organize ourselves globally around these vital roles, the easier it will be for us to solve the nature and climate crisis.

This begins with recognizing that humans belong exactly as we are on Planet B, even though there is a steady drumbeat of negativity on Planet A telling us that if humans disappeared off the face of the Earth, the planet would be fine.

During the Vegan Convergence Of the Peoples #20 (V-COP20), the 20th edition of a quarterly conference series that Climate Healers has conducted online to evolve from Planet A into Planet B, Capt. Paul Watson of the eponymous Captain Paul Watson foundation and co-founder of Greenpeace and Sea Shepherd, made a startling revelation about a new children's book that he's writing, called Starship Earth. Capt. Watson said[2],

> *"Starship Earth needs engineers and we have them. These engineers keep everything running, maintain all of the systems. We humans, we are not engineers. We are passengers. We are having a great time amusing ourselves, entertaining ourselves. But what we are doing is we are killing the engineers, we are committing murder on all of these*

engineers. We are wiping them out, the bees, and the trees, and the fish, and the microbes, and the fungi..."

In that telling, Starship Earth would be better off without humans. During a talk in Mysuru, Karnataka, India, I asked the audience if they thought the planet would be better off without humans. And to my horror, almost every Conscious Vegan in the audience did.

When I first watched the documentary *Earthlings* in early 2009, I too was ashamed to be a member of my species and thought the same way about humans. Therefore, I'm not surprised that Conscious Vegans who have informed themselves on the cruel reality of animal husbandry are not too enamored of our species. However, as Dr. Baruch[3] pointed out at V-COP20, it is extremely important that we overcome that self-loathing and begin to love ourselves and our species to become effective change makers.

The Climate Healers story is that humans are also engineers, not just passengers, on Starship Earth. We are doing the difficult job of installing a climate control system so that Starship Earth has the optimum conditions for life to thrive.

Yes, we have killed a lot of our fellow engineers and made a mess in the process, but we can now clean up and make amends to our fellow engineers so that Starship Earth will benefit in the long run.

It is important that we tell such a story to children so that they know and feel that they belong. It is equally important that all Conscious Vegans internalize such a story so that we know and feel that we belong. We must also know that humans belong as a species because Nature is the perfect system design.

It is only when we know and feel that we belong that we can truly shine our light and help others come out of their closet.

Shining our light means seeing and highlighting the positive in people and especially, the news.

Whatever we focus on and highlight amplifies in the universe.

There IS a Planet B

If we keep seeing darkness everywhere and highlight that, everyone around us notices that darkness as well and it gets amplified.

If we keep seeing the light everywhere and highlight that, everyone around us notices that light as well and it gets amplified.

Indeed, in the following chapters, we will make the case that the planet **would not be fine** if humans disappeared off the face of the Earth and that the VEGAN thermostat species is a vital role necessary for Mother Earth to thrive well into the future.

We must recognize that the seemingly destructive activities of our ancestors had some ecological benefit and the Earth would fare much better if our generation and our descendants completed the job that our ancestors started and had been working on for at least 50,000 years if not for 500,000 years.

---

[1] This statement is taken from *Codes for a Healthy Earth*. Link https://www.codes.earth/thecodes accessed on March 7, 2025. Please see also, Chapter 27 ibid.

[2] 2. Please see the video embedded in https://trello.com/c/1GO0fmOg/1308-keynote-rewilding-the-ocean. Link accessed on March 7, 2025.

[3] Please see the video embedded in https://trello.com/c/7XaxfFA6/1315-balancing-loving-you-the-micro-as-a-precursor-to-balancing-loving-earth-the-macro. Link accessed on March 7, 2025.

11. Setting Context in Time

"The Earth is not part of the human story.
The human story is part of the Earth's
story...
The problem is we have tried to tell the human story
without telling the Earth's story."
– Father Thomas Berry

We begin with the premise that humans are a part of Nature and have never been apart from Nature. There is considerable scientific evidence for the critical role we're playing in Nature, based on paleoclimatology. Just like the elephant breaking branches off trees, we can see a pattern that shows how humans have been engaged in one of Nature's long-term projects without being fully aware of our role in it.

In many faith and wisdom traditions, for example in Yoga, it is told that our separation from Nature is a delusion, the greatest delusion of them all. It is a delusion in the same sense as my belief that there was no fire in the house when I was on the phone with my colleague was a delusion, since there was indeed a fire raging in the garage.

Just because we wash our food, expel the soil and dirt from inside our homes, and fantasize endlessly in science fiction movies about traveling to distant galaxies does not make us separate from Nature.

There IS a Planet B

We are firmly a part of Nature even on Planet A, whether we realize it or not.

In order to understand and appreciate our role in Nature, it is important to set the context in time. Given all the commotion over global warming on Earth, it may be surprising to hear that for the past 33.9 Million years, the Earth has been in the Sixth Major Ice age in its history, the late Cenozoic Ice age[1].

An ice age is considered major when there is evidence of significant glaciation over millions of years leaving a clear mark in the fossil record. The previous major ice age occurred 300 Million years ago, also when atmospheric CO_2 levels were quite low.

Our sun is a G-type main sequence star (G2V) which increases in intensity over time. If the sun had behaved like every normal G2V star, then the sun's intensity must have increased by 1% for every 100 million years of its existence. As the sun got warmer and warmer over time, Mother Earth began shedding her greenhouse gas blankets as needed in order to regulate the conditions for life in the biosphere.

This regulation is not very precise and requires time constants on the order of millennia through rock weathering and similar processes. Species also evolve thereby altering the optimum climatic conditions for the well being of life in the biosphere at any given time and consequently, the optimum greenhouse gas concentrations in the atmosphere.

As greenhouse gas concentrations in the atmosphere got reduced, the Earth became extremely sensitive to small changes in the amount of sunlight falling on Earth. The Earth is the only planet in our solar system that harbors life within the Habitable Zone (HZ) around the Sun.

The HZ is the zone where liquid water would be available on the surface of the planet so that life can thrive. The boundaries of the HZ are typically calculated using 1-Dimensional cloud free climate models as pioneered by Kasting et al[2]. Using this model, they showed that the HZ around our sun, at present, is 0.95AU to

1.67AU, where AU is the Astronomical Unit, the average distance of the Earth from the Sun. In this model, the Earth is presently 0.05AU away from the inner edge of the Habitable Zone around the sun.

A more refined 1-Dimensional radiative-convective, cloud free climate model, with updated H_2O and CO_2 absorption coefficients was used by Dr. Ravi Kopparapu and others at Penn State University[3] to show that the HZ limits of our solar system is from 0.99AU to 1.70AU. This means that the Earth is **extremely close** to the inner edge of the Habitable Zone, just 0.01AU from the inner edge, and therefore, very sensitive to small perturbations in the greenhouse gas composition in the atmosphere.

On Venus, which is orbiting at a distance of 0.72 AU from the Sun, the average surface temperature is estimated to be 462°C. On Mars, which is orbiting at a distance of 1.57 AU from the Sun, also within the Habitable Zone around the Sun, the average surface temperature is -60°C. As one might expect, the closer the planet is to the Sun, the hotter it gets.

Likewise, the closer the Earth is to the inner edge of the HZ of the Sun, the tendency is to get too hot. Paradoxically, the major Late Cenozoic ice age is occurring as the Earth got closer to the inner edge of the HZ around the Sun. This is because when atmospheric greenhouse gas concentrations are lowered, small changes in their concentrations cause large changes in the amount of heat retained on Earth and the coarsely controlled, passive feedback mechanisms on Earth are no longer capable of regulating the surface temperature adequately.

Indeed, the concentration of the main greenhouse gas in the atmosphere that regulates the surface temperature of the Earth, CO_2, had been fluctuating between 180ppm and 300ppm over the past 3 million years. These are historically low levels compared to the estimate that the CO_2 concentration in the atmosphere was between 1700ppm and 7500ppm, 400 million years ago when the sun was presumably 4% cooler than it is today.

There IS a Planet B

Both water vapor and greenhouse gases such as CO_2 are responsible for the Earth's atmosphere to warm. Think of CO_2 as the glass roof of a greenhouse and water vapor as the evaporated moisture from the soil within the greenhouse. Without the glass trapping the heat, the moisture in the soil would not be heating up the atmosphere in the greenhouse as much.

Likewise, without the greenhouse gas molecules in the atmosphere trapping certain wavelengths of the electromagnetic spectrum in the infrared region, water vapor would not be as available to warm up the atmosphere. The greenhouse gases act as catalysts to assist the water vapor to warm up the atmosphere.

The Late Cenozoic ice age is the first ice age to occur during the era of the mammals and during this ice age, for the past 3 million years, the earth has been going in and out of glacial periods, following what are known as Milankovitch cycles. These cycles are caused by small wobbles in the Earth's orbit around the sun which result in slight variations in the amount of solar energy falling on Earth. These slight variations are sufficient to cause the Earth to swing from a cold glacial period to a warm interglacial period and back again.

Over the past million years, these warm interglacial periods occurred once every 100,000 years or so for a short duration of about 10,000 years or less, as part of what is known as the Quaternary Ice Age.

---

[1] Jansen, E. et al., 2007: Palaeoclimate. In: Climate Change 2007: The Physical Science Basis. Contribution of Working Group I to the Fourth Assessment Report of the Intergovernmental Panel on Climate Change [Solomon, S., et al. (eds.)]. Cambridge University Press, Cambridge, United Kingdom and New York, NY, USA. Link https://www.ipcc.ch/site/assets/uploads/2018/02/ar4-wg1-chapter6-1.pdf accessed on March 7, 2025.

[2] Kasting, J. F. et. al, "Habitable Zones Around Main Sequence Stars," ICARUS 101, 108-128, 1993. Link https://www.researchgate.net/publication/11809380_Habitable_zones_around_main_sequence_stars accessed on March 7, 2025.

[3] Kopparapu, R. et al, "Habitable Zones Around Main-Sequence Stars: New Estimates," Earth and Planetary Astrophysics, 2013, Doi: 10.1088/0004-637X/765/2/131, https://arxiv.org/abs/1301.6674 accessed on March 7, 2025.

12. The Quaternary Ice Age

"Truth is stranger than fiction, but it's because fiction is obliged to stick to possibilities. Truth isn't." – Mark Twain

During the Quaternary Ice Age, the atmospheric CO_2 levels have been varying between 180 and 300 parts per million, which are unprecedentedly low levels. The earth has been swinging between warm interglacial periods and cold glacial periods with a range of 6-8°C difference in the Earth's global surface temperature between the warm and cold periods.

Paradoxically, when the sun got warmer and warmer and pushed the Earth towards the inner edge of the Habitable Zone, the Earth's surface temperature, for the most part, became too cold mainly because the regulating mechanisms for maintaining a stable climate on Earth became too coarse to respond adequately to the changes in the environmental conditions.

Within the biosphere, these wild swings in the global surface temperature and climatic conditions were not optimum for the well-being of life. Events like the Dansgaard-Oeschger cycles[1] with rapid warming of 5-8°C within decades to centuries and the Heinrich events[2] with ice sheet discharges leading to rapid cooling occurred frequently during the Quaternary Ice Age.

There IS a Planet B

During this cataclysmic period for the well-being of life, the Earth spawned us as a species with ordinary sensory skills, but with the extraordinary biological features of opposable thumbs combined with a prefrontal cortex. As in the past when life faced environmental challenges, life is once again seeking to increase in complexity in order to overcome the challenges.

Humans are an integral part of that increase in complexity. We don't smell too well, hear too well, see too well or even climb trees too well. Therefore, we were easy to catch for the fearsome predators that were chasing prey a million years ago, which forced our ancestors to use their brains to learn how to control fire at least 500,000 years ago[3] and use their opposable thumbs to create weapons to defend themselves.

Then, about 50,000 years ago, during the previous glacial period, we formed partnerships with wolves, who became our dogs and lent us their superior senses of smell, sight and hearing so that together we could survive during our migration out of Africa and into every part of the world.

During the current warm period, the Holocene era, the reconstructed temperature of the Earth initially closely matched the temperature that was seen during the warm interglacial period three glaciations ago, except that instead of going down to another glacial period in about 5,000 years, the temperature stayed fairly constant over the past 10,000 years and then increased lately by 1.3°C with respect to pre-industrial levels.

---

[1] Schmidt, M. W. and Hertzberg, J. E., *Abrupt Climate Change During the Last Ice Age*, Nature Education Knowledge 3, 2011. Link https://www.nature.com/scitable/knowledge/library/abrupt-climate-change-during-the-last-ice-24288097/ accessed on March 7, 2025.

[2] Bassis, J. N., et. al, *Heinrich Events Triggered by Ocean Forcing and Modulated by Isostatic Adjustment*, Nature, vol. 542, 2017. Link https://www.nature.com/articles/nature21069 accessed on March 7, 2025.

[3] Gowlett, J. A. J, *The Discovery of Fire by Humans: A Long and Convoluted Process*, Philosophical Transactions of the Royal Society B, Biological Sciences, Jun 2016. Link https://royalsocietypublishing.org/doi/10.1098/rstb.2015.0164 accessed on March 7, 2025.

13. How did we Heat the Climate?

"You are not a drop in the ocean. You are the entire ocean in a drop"
— Rumi

When we examine what happened to keep the global surface temperature constant during the Holocene era, we see the unmistakable fingerprint of our human ancestors. The atmospheric CO_2 level during this Holocene era closely matched what was happening during the three previous interglacial periods until about 6,000 years ago and then it deviated significantly and stayed essentially constant at 280 ppm instead of going down to 180 ppm.

Likewise, the methane level in the atmosphere closely matched what was happening during the three previous interglacial periods until about 4,000 years ago and then it deviated significantly and increased sharply.

The Early Anthropocene Hypothesis of William Ruddiman[1] proposes that these deviations were due to the extensive land clearing for agriculture that our ancestors engaged in during the agricultural revolution, along with the rice cultivation and the domestication of ruminants that contributed to the methane emissions. When these greenhouse gas levels were held steady in the atmosphere, the temperature of the earth stabilized during the Holocene era, creating the conditions for agriculture to flourish.

There IS a Planet B

Since William Ruddiman proposed the Early Anthropocene Hypothesis in 2003, he has been addressing anomalies from skeptical scientists trying to disprove his hypothesis. Thus far, he has succeeded in explaining all the anomalies and the hypothesis still stands.

The temperature of the Earth rose above the glaciation threshold 11,700 years ago and it would have returned below the glaciation threshold about 5,000 years ago, except that our ancestors kept the temperature constant through these land use change activities. Then, 250 years ago, we discovered fossil fuels and cranked it up a notch. We also began raising animals at industrial scales and the result is that the temperature of the earth has increased by about 1.3°C from pre-industrial levels today.

Now, we are at an inflection point, with three possible futures ahead of us:

1. We can continue heating up the climate so that the earth's surface temperature increases to 3-4°C above pre-industrial levels and causes both biological annihilation and runaway climate change, leading to the end of life on Earth;
2. We can continue heating up the climate so that humans go extinct in the near term fast enough for the Earth to return back to another Quaternary glaciation; or
3. We consciously start cooling the climate and restoring the lost biodiversity of the planet, eventually leading to the stabilization of the earth's temperature at pre-industrial levels.

Of the three options before us, we will explore the only life affirming option #3 in detail. We will assume that **Mother Earth needs a finely controlled, active feedback climate regulating mechanism**, which humans, as a technologically adept species, are now well equipped to provide.

[1] William Ruddiman's book, *Plows, Plagues and Petroleum: How Humans Took Control of the Climate*, Princeton University Press, 2006, has sparked lively scientific debate since it was first published—arguing that humans have actually been changing the climate for some 8,000 years—as a result of the earlier discovery of agriculture. His Early Anthropocene hypothesis has withstood scrutiny for two decades and deserves to be taken seriously as a theory, not just a hypothesis. Link https://press.princeton.edu/books/paperback/9780691173214/plows-plagues-and-petroleum accessed on March 7, 2025.

14. How we Use The Earth

"This we know: the Earth does not belong
to man — man belongs to the Earth."
– Ted Perry

On Planet A, humans are led to believe that the Earth is here for humans to use. Over the past ten thousand years, use the Earth we did. There is some dispute over whether the lakes, rivers, vegetation and lush green forests of North Africa disappeared and turned into sandy desert due to natural climate variations causing humans to resort to animal husbandry in order to survive or whether deforestation for animal husbandry caused the drying of the North African climate[1]. But there is no dispute that the subsequent expansion of the Saharan desert to extend all the way into China as the Gobi desert and all the way into India as the Thar desert was primarily due to animal husbandry. The human fingerprint on this desertification is evident today as the UN Convention to Combat Desertification has been documenting for the past three decades[2].

Fully 19% of the 13 Billion hectares of the ice-free land area of the planet is such deserts, mountains and other barren land with minimal human use[3].

We tend to think of ourselves as being so numerous that we are over running the Earth. Indeed, humans comprise 36% of the biomass of

82

There IS a Planet B

all mammals on Earth[4], with our farmed animals making up 60% of the biomass and wild animals, from mice to elephants to whales, comprising the remaining 4%. If we consider just land mammals, wild animals comprise just 2% of the biomass of mammals on land, with humans comprising 37% and farmed animals 61%.

However, if all humans on Earth were to do a group hug in one spot, we would just about fill up one large city the size of New Delhi, India and that would be it. This is why all the cities, railroads, highways, airports, all the built land cover only about 1% of the ice-free land area of the planet.

It is not our biomass that matters, but what we consume. In order to meet our consumer demand for forest products, managed forests, mono-cultured for timber, paper, etc., make up 22% of the ice-free land area of the planet.

Original forests where wild animals still live, comprise just 9% of the ice-free land area of the planet. This is also the land where indigenous people live and the land that is being appropriated for more mining and meat and dairy production in order to grow the global human economy.

By far, the single biggest use of the Earth today is to procure food for human consumption. We eat about 1.59 Giga tons (Gt) of food in terms of dry weight[5], but in order to provide that food on our plates, we are extracting 9.05 Gt of food from the planet. To calculate the dry weight, we remove the water from the food and weigh it in order to make an apples to apples comparison between different food options.

There is a nearly 6 to 1 reduction between the food we procure and the food we eat because most of the food goes to feed our farmed animals.

37% of the ice-free land is used for grazing farmed animals and if we include the fact that half the cropland output goes to feed animals, we are using about 43% of the ice-free land area of the planet to raise farmed animals. These animals provide us with just 12% of the food we eat, in terms of dry weight. Animal foods are so

inefficient because we have to feed our animals 39 kgs of food in order to get 1 kg of the meat and dairy we consume, in terms of dry weight.

We are growing the plant foods we eat today, the fruits, vegetables, grains, nuts and seeds in about 6% of the land area of the planet. This is providing us with 85% of the food we eat, in terms of dry weight.

The remaining 3% of the food we eat comes from the ocean, for which we are bottom trawling an area the size of South America each year in order to catch the last remaining fish in the ocean.

When we all go vegan and increase the current 85% of plant foods we eat towards 100%, we can return 40% of the ice-free land area of the planet back to Nature, as well as the entire ocean, effectively releasing a total of 80% of the earth's surface.

Of course, Veganism isn't just about the food we eat, but also about not using animals for the clothing we wear and the housing we live in and so on. It is only when we go vegan that we get the full release of the land used for raising farmed animals. If we eat only plant foods, but insist on wearing leather jackets and boots made from animal skins, then the industrial sector is perfectly capable of raising animals for just their skins. Indeed, it already does raise some animal species for just their skins.

Today, mainly to accommodate animal husbandry, humans have cut down half the trees on the planet, from 5-6 trillion that existed 10,000 years ago down to 3 trillion[6]. But the remaining 3 trillion trees and the soil they live on are storing twice as much carbon as in the entire atmosphere and four times as much carbon as in all the fossil fuels we have burned to date[7]. When we go vegan and restore most of the missing 3 trillion trees on 40% of the ice-free land area of the planet that gets freed up from animal husbandry, we can literally reverse the post-industrial increase in the atmospheric CO_2 level and thereby heal the climate.

There IS a Planet B

[1] Wright, D. K., *Humans as Agents in Termination of the African Humid Period*, Frontiers in Earth Science, Sec. Quaternary Science, Geomorphology and Paleoenvironment, vol. 5, Jan 2017. Link https://www.frontiersin.org/journals/earth-science/articles/10.3389/feart.2017.00004/full#B10 accessed on March 7, 2025.

[2] The Global Land Outlook (GLO) is the flagship publication of the UN Convention to Combat Desertification (UNCCD) and it underscores land system challenges, showcases policies and practices, and points to pathways to scale up sustainable land and water management. GLO details how much animal agriculture has contributed to land degradation. Find more here: https://www.unccd.int/resources/global-land-outlook/overview, accessed on March 7, 2025.

[3] The land use breakdown described in this chapter is taken from the UN IPCC Climate Change and Land Special Report of 2019. Arneth, A., et. al, *IPCC Special Report on Climate Change, Desertification, Land Degradation, Sustainable Land Management, Food Security, and Greenhouse gas fluxes in Terrestrial Ecosystems*, UN Intergovernmental Panel on Climate Change, Aug 2019. Link: https://www.ipcc.ch/site/assets/uploads/2019/08/Fullreport-1.pdf accessed on March 7, 2025.

[4] Bar-on et. al, *The Biomass Distribution on Earth*, Proceedings of the National Academies of Sciences, Biological Sciences, vol. 115, no. 25, 2018. Link https://www.pnas.org/doi/full/10.1073/pnas.1711842115 accessed on March 7, 2025.

[5] Smith, P., et. al, *Agriculture, Forestry and Other Land Use (AFOLU)*. In: Climate Change 2014: Mitigation of Climate Change. Contribution of Working Group III to the Fifth Assessment Report of the Intergovernmental Panel on Climate Change, Cambridge University Press, Cambridge, United Kingdom and New York, NY, USA, 2014. Link https://www.ipcc.ch/site/assets/uploads/2018/02/ipcc_wg3_ar5_chapter11.pdf accessed on March 7, 2025.

[6] Crowther, T.W., et. al, *Mapping Tree Density at a Global Scale*, Nature 525, Sep 2015, pp. 201-205. Link https://www.nature.com/articles/nature14967 accessed on March 7, 2025.

---

[7] Canadell, P., et a., *Global Carbon and Other Biogeochemical Cycles and Feedbacks* in Climate Change 2021 – The Physical Science Basis: Working Group I Contribution to the Sixth Assessment Report of the Intergovernmental Panel on Climate Change, Cambridge: Cambridge University Press; 2021. Link https://www.ipcc.ch/report/ar6/wg1/downloads/figures/IPCC_AR6_WGI_Figure_5_12.png accessed on March 7, 2025.

15. The VEGAN Rewilding Solution

*"We should all be eating fruits and vegetables as if
our lives depend on it — because it does."*
– Dr. Michael Greger

Going vegan may be just one small step for each of us, but one giant leap for all life on earth. Due to the enormous impact on the life-support systems of the planet, using animals for food, clothing or any other purpose is not a personal choice, but an intensely political choice.

When we choose to go vegan and rewild the planet on the land that gets freed up from animal husbandry and rewild the ocean, not only can we heal the climate, the fourth worst planetary boundary transgression, but we can address all the other transgressions that the Stockholm Resilience Center identified in 2023.

This is why the Economic Survey of the Finance Ministry of India[1] wrote that *"nudging family, friends and colleagues to a more sustainable dietary preference and moderation in lifestyles globally may be an idea whose time has come."*

However, this is not just about dietary preferences but also about our use of animals for any purpose whatsoever. If we stop eating animal foods, but continue to wear animals, the industry is perfectly

There IS a Planet B

capable of raising animals just for their skins and therefore, it won't result in land getting freed up from animal husbandry. In fact, the industry already raises a number of animals, minks, rabbits, etc., for just their skins.

Eliminating animal use requires nudging people on their diets as well as their other consumer preferences. The picky vegan who looks through the list of ingredients in products in order to weed out any use of animals is sending a signal to the industry sector that any use of animals as resources is not acceptable. It is this signal that can eventually free up land from animal husbandry.

Of all the planetary boundary transgressions, the least violated transgression is fresh-water change[2]. When we stop farming animals and restore native ecosystems on 40% of the land, we will also restore the water cycles of the planet and reverse this transgression.

The next is land systems change, which is solved when we stop farming animals and return 40% of the land back to Nature.

The next is climate change which is resolved when we restore the missing 3 trillion trees on the land that gets freed up from animal husbandry.

The next is biogeochemical flows or nitrogen and phosphorous loading. This is caused by using fertilizers on our cropland. Since half the crops are used to feed animals, this transgression goes from red towards green when we stop farming animals and go vegan.

The next is novel entities or chemical pollution, which would be safely stored away in regenerating forests when we stop farming animals and go vegan. Eating animal foods currently delivers concentrated doses of this chemical pollution into our bodies through bioaccumulation. Therefore, going vegan addresses chemical pollution for both the earth and for ourselves.

All of these transgressions impact wildlife, and biosphere integrity is the worst of the six planetary-boundary transgressions. By freeing the land from farmed animals to restore habitats for wild animals

and allowing them to live freely in the ocean, we will resolve this transgression as well. If instead, we let wild animals die off, we will also die off.

---

[1] 1. The Economic Survey of India is published by the Chief Economic Advisor of India, Dr. V. Anantha Nageswaran and the quote is from Chapter 10 of the Economic Survey: Climate and Environment: Adaptation Matters. Link https://www.indiabudget.gov.in/economicsurvey/doc/eschapter/echap10.pdf accessed March 4, 2025.

[2] Richardson, K., et al., *Earth Beyond Six of Nine Planetary Boundaries,* Science Advances, vol. 9, no. 37, Sep 2023. Link https://www.science.org/doi/10.1126/sciadv.adh2458 accessed March 6, 2025.

16. The Animal Agriculture Position Paper

"Even if you are a minority of one, the truth is the truth."
– Mahatma Gandhi

Foundational Myth on Planet A:
Fossil fuel use is the leading cause of climate change.

Foundational Truth on Planet B:
Animal husbandry is the leading cause of climate change, while
stopping fossil fuel use now is suicidal.

I wrote a position paper in 2019[1] showing that when we take into
account the potential carbon absorption of the land used for animal
husbandry as well as all of the greenhouse gas emissions caused by
that sector, animal husbandry is responsible for at least 87% of
greenhouse gas emissions on an annual basis.

Based on the Sixth Assessment Report of the UN IPCC issued in late
2021, I was able to update that lower bound to **118%**.

Now you may wonder — how can a single activity be responsible
for greater than 100% of the total? It turns out that when we all go
vegan, but continue our other activities as is, drive around in cars,
fly in planes, heat and cool our homes as we do today, we will still

There IS a Planet B

see the CO_2 levels in the atmosphere decrease year by year. When we all go vegan, we are letting the planet rewild on 40% of the land area of the planet as well as the entire ocean so that 80% of the earth's surface gets returned back to Nature.

It is true that on some areas of the land and in parts of the ocean, humans will have to intervene to help restore the original ecosystems in that area by cleaning up toxic wastes, restoring water retention structures, planting native vegetation, introducing native animal species, etc. This can be accomplished when the primary objective of humanity switches towards the restoration of Nature instead of merely extracting from Nature. This happens when we formally recognize that we have already extracted enough for our larger objective of stabilizing the climate and the life-support systems of the planet.

In systems engineering, an opportunity cost is always calculated assuming that we are assigning reasonable resources and effort for the alternate path being considered. In the position paper, the opportunity cost of the land used for animal husbandry was calculated assuming that Nature is able to sequester at least as much CO_2 as that embedded in all the food that our farmed animals are consuming today when we are consciously restoring the ecosystems of the planet. In addition, we assumed that the vegetation that our farmed animals are consuming are all above ground, while a commensurate amount of CO_2 sequestration occurs below ground in root systems and the soil that the vegetation is thriving on.

One of the key areas in which our position paper differs with the UN IPCC's approach is in the accounting of land use change emissions. The IPCC uses *net* accounting for land use change emissions whereby emissions from land that gets deforested is offset with regrowth on managed or abandoned land.

The IPCC counts deforestation emissions as if the land has been turned from forest into desert. Of course, it is very hard to turn forest into desert as the deforested land attempts to regenerate every year. Then, the IPCC assumes that all vegetation that is eaten by animals, cut down and burnt on that land, year after year, does

not need to be counted since it already counted all the emissions in the first year itself.

All of those annual emissions become freely available for the use of animal husbandry. After a certain number of years, the land gets tired of regenerating and eventually becomes a desert. However, as far as the IPCC is concerned, the emissions from that land is exactly the same regardless of whether Nature attempted to regenerate on that land for 2 years, or 10 years or 50 years before giving up.

In the position paper, we counted some of these recurring emissions as best as we could, following the work of Goodland and Anhang[2]. Now, Gerard Wedderburn-Bisshop[3] has documented that switching from *"net"* accounting to *"gross"* accounting for land use change emissions alone increases its contribution to overall CO_2 emissions by nearly a factor of 3. In *gross* accounting, we count all of the annual deforestation emissions instead of offsetting a majority of the emissions against regrowth on managed and abandoned land.

When we consider photosynthesis as a gift of Nature, then even the annual emissions from animal feed, pasture maintenance fires, bottom trawling of the ocean, etc., would need to be counted as humans are responsible for *all* of these emissions, while humans ought not to be taking credit for *any* of the photosynthesis that is occurring to offset these emissions.

Just as there is an opportunity cost for the land used for animal husbandry, there is an *opportunity benefit* for fossil fuel use in the form of cooling gases that are masking a significant portion of the global warming. There is considerable ambiguity on the cooling impact of SO_2 due to cloud effects which are offset by the warming cloud effects from CO_2 emissions.

The UN IPCC is assuming 3°C for the equilibrium temperature rise on the planet for CO_2 doubling in the atmosphere so that the CO_2 added to the atmosphere since 1750 has contributed 2.06 W/m^2 of ERF until 2020. In 2020, the International Maritime Organization passed a ruling[4] limiting the sulphur content of the fuel that can be burned over the open seas. Hansen et al. have made a compelling

There IS a Planet B

case that this ruling was responsible for the two hottest years in recorded history, 2023 and 2024.

According to Hansen et. al[5], this unexpected increase in global surface temperature during 2023 and 2024 can be explained in models if the equilibrium temperature rise for a doubling of the CO_2 concentration in the atmosphere is 4.5°C and not 3°C. Correspondingly, the ERF of CO_2 increases from 2.06 W/m^2 to 3.09 W/m^2 and the ERF of SO_2 changes from -0.94 W/m^2 to -1.97 W/m^2.

As a result, stopping fossil fuel use abruptly would increase the global warming ERF on the planet by 2.3 W/m^2, not just 1.3 W/m^2. This puts the UN IPCC in the unenviable position of arguing that fossil fuel use is the leading cause of climate change, provided:

1. we don't count the vast majority of the CO_2 emissions from animal husbandry;
2. we don't include the opportunity cost of the land used for animal husbandry;
3. we undercount the impact of methane emissions by a factor of 3; and
4. we ignore the cooling effects of fossil fuel use that nearly completely offsets its warming effects.

Surely we can do better, more compassionate science on a nature and climate crisis of existential proportions on Planet B.

---

[1] Rao, S., *Animal Agriculture is the Leading Cause of Climate Change: A Position Paper*, Journal of Ecological Society, vol 32-33, 2021. The paper is also hosted on the Climate Healers website here: https://climatehealers.org/the-science/animal-agriculture-position-paper/ Link accessed on March 7, 2025.

[2] Goodland, R. and Anhang, J. M., *Livestock and Climate Change: What if the Key Actors in Climate Change were Pigs, Chickens and Cows*, World Watch, November-December 2009, Worldwatch Institute, Washington DC, USA, pp. 10-19. The link https://awellfedworld.org/wp-content/uploads/Livestock-Climate-Change-Anhang-Goodland.pdf was accessed on March 7, 2025.

[3] Gerard Wedderburn-Bisshop, *Increased Transparency in Accounting Conventions Could Benefit Climate Policy,* Environmental Research Letters, vol. 20, no. 2, Feb 2025. Link https://iopscience.iop.org/article/10.1088/1748-9326/adb7f2 accessed on March 7, 2025.

[4] The International Maritime Organization (IMO) reduced the allowable sulphur content on fuels burned over the open seas from 3.5% to 0.5%, starting in 2020. Change of global aerosol forcing from this limit on ship emissions, based on IPCC's formulation of aerosol forcing, is calculated as 0.079 W/m2 in Z. Hausfather and P. Forster, *Analysis: How Low-sulphur Shipping Rules are Affecting Global Warming,* Carbon Brief, (3 July 2023). Link https://www.carbonbrief.org/analysis-how-low-sulphur-shipping-rules-are-affecting-global-warming/ accessed on March 7, 2025.

[5] In this paper, Hansen et al. calculate the change of global aerosol forcing from the IMO's restriction on the sulphur content of fuel for open seas shipping to be 0.5 W/m2, not 0.079 W/m2. Hansen JE, et al. *Global Warming Has Accelerated: Are the United Nations and the Public Well- Informed?* Environment: Science and Policy for Sustainable Development. 2025 Jan 2;67(1):6–44. Link https://www.tandfonline.com/doi/full/10.1080/00139157.2025.2434494#d1e751 accessed March 6, 2025.

17. The Climate Bathtub Model

"All models are wrong,
but some are useful."
– George Box

The Climate Bathtub model is a realistic system model to help us make the right choices on the climate crisis.

Imagine that a baby representing all life on earth is stuck in a "Climate Bathtub" filling up with water at 50 Liters per minute (L/m) from two running faucets — the Burning Machine faucet and the Killing Machine faucet. When you arrive, the bathtub has filled up with 1100 Liters (L) of water.

The Burning Machine (BM) faucet is pouring 25 L/m into the bathtub and the faucet is connected to the overhead Aerosols cistern containing 500 L of water. For every 1 L/m that you turn down the BM faucet, it lets 20 L of water pour out of the overhead cistern into the bathtub instantly. By the time you turn off the Burning Machine faucet completely, all 500 L from the Aerosols cistern will be added to the Climate Bathtub.

The Killing Machine (KM) faucet is also pouring 25 L/m into the bathtub, but the faucet is connected to the drain of the bathtub and the underground Vegan Reforestation (VR) tank.

There IS a Planet B

As the KM faucet is turned off, it proportionately opens up the drain and lets the water in the bathtub drain into the VR tank. The maximum capacity of the drain is 35 L/m.

The VR tank can hold 2000 L of water. The baby is already struggling to breathe and will most certainly drown if the climate bathtub holds more than 1200 L of water.

Questions:

1. As a responsible decision maker of this planet who has been given the power to turn off the Burning Machine and Killing Machine faucets at will, can you save the baby and drain the bathtub without overflowing the VR tank?

2. If the answer to 1 is YES,
   a. Determine how long we can procrastinate before it becomes impossible to save the baby and drain the bathtub without overflowing the VR tank.
   b. Determine the minimum we need to raise the water level in the bathtub in order to save the baby and drain the bathtub and how long would it take to do so.

3. Answer Questions 1&2, assuming the objective is to lower the water level in the bathtub to 500 L instead of draining it completely.

The 1100 L of water in the Climate Bathtub corresponds to the 1100 Gt CO_2 that we have added to the atmosphere since 1750 in the industrial era. The inflow of 50 L/m into the bathtub from the two faucets corresponds to the roughly 50 Gt CO_2 that we add to the atmosphere every year. I have split human activities into two machines so that the Burning Machine includes the fossil fuel and industry sectors, while the Killing Machine includes animal agriculture, forestry and waste.

Therefore, 1 L in the bathtub corresponds to 1 Gt CO_2 in the atmosphere and 1 min corresponds to 1 year.

When the KM faucet is turned off completely, it reduces the inflow into the bathtub by 60 L/m, which is 120% of the inflow when the faucet is on completely. This corresponds to the 118% reduction in greenhouse gas emissions that we expect when we stop animal husbandry completely.

The 2000L capacity of the VR tank corresponds to the estimated 2000 Gt CO_2 that can be stored through rewilding grazing lands.

The 500L of water in the Aerosols cistern corresponds to the nearly 50% increase in ERF that we can expect if we phase down fossil fuel use immediately.

At the moment, the answer to Question 1 is YES and it is possible to save the baby and drain the bathtub without increasing the water level in the bathtub above 1100 L.

The closed form solution to Question 2b is
- turn off the KM faucet immediately, and
- turn off the BM faucet gradually according to the equation:
  $$BM(t) = 35 - 10\exp(t/20) \text{ for } 0 < t < 20\ln(3.5) = 25.06 \text{ mins}$$
  $$= 0 \text{ for } t \geq 25.06 \text{ mins}$$

This would keep the bathtub level at 1100L until t = 25.06 mins and the bathtub would be drained at t = 56.49 mins at which point the VR tank would hold 1977 L of water.

The answer to Question 2a is 23/50 = 0.46 mins.

I'll leave answering Question 3 as an exercise to the reader.

There IS a Planet B

Of course, the Climate Bathtub model makes it very clear that if we turn off the Burning Machine faucet precipitously while keeping the Killing Machine faucet fully turned on as mainstream climate spokespeople are recommending on Planet A, the baby would drown instantly.

18. The Need for a Spiritual and Cultural Transformation

"Science is not only compatible with spirituality —
it is a profound source of spirituality."
– Carl Sagan

Animal husbandry is the leading cause of not only climate change, but also biodiversity loss and ecosystem collapse, the other two major environmental problems identified by the UN at the Rio summit in 1992 and ending animal husbandry is the leading solution to address all six planetary boundary transgressions. However, this is not common knowledge because as Gus Speth, the cofounder of Natural Resources Defense Council (NRDC) and the former Dean of Environmental Studies at Yale University, was quoted in a BBC Radio 4 program in 2013[1]:

"I used to think the top environmental problems were biodiversity loss, ecosystem collapse and climate change. I thought that with 30 years of good science we could address these problems.

But I was wrong. The top environmental problems are selfishness, greed and apathy and to deal with those, we need a spiritual and cultural transformation, and we scientists don't know how to do that."

There IS a Planet B

This spiritual and cultural transformation must come from the people themselves. Global hunger is a choice we are making at least three times a day, ecological destruction is a choice we are making at least three times a day. And we can choose to go vegan, starting today, once we see through the falsehoods underpinning our choices.

Systems based on fear are dictated from the top down, but systems based on love are originated from the bottom up through conscious choices. Love comes with the freedom to choose. Even in the Bhagavad Gita, God tells Man[2], *"I love you so much that I set you free even from me, and you are free to choose."*

While hard scientists may not know how to do a spiritual and cultural transformation, social scientists do have a good idea. As a systems engineer, I can show that selfishness, greed and apathy are being promoted systematically on Planet A, since selfishness and greed are handsomely rewarded in the economic system and the public is rendered apathetic with mass advertising that is intended to make them feel incomplete so that they become compliant consumers. When we are constantly told to measure up against an impossible ideal, it is understandable that a majority of us concede the battle and give up on Planet A.

---

[1] Quoted in BBC Radio 4, Shared Planet: Religion and Nature. Link https://www.bbc.co.uk/sounds/play/b03bqws7 accessed on March 7, 2025.

[2] The Bhagavad Gita is Hinduism's core sacred text, a dialogue of and on Dharma, right action. It enshrines the essential values of the Vedic, Upanishadic and epic traditions - shruti (the revealed) and smriti (the remembered). Its structure is informal questions and answers; its mode is enquiry and search; its goal is self-discovery and spiritual illumination.
Please see Canto XVIII, Verses 63:64 in
https://www.abebooks.com/9788174363244/Bhagavad-Gita-P-Lal-8174363246/plp, link accessed on March 7, 2025.

19. From Climate *Heating* to Climate *Healing*

"Adding wings to caterpillars does not create butterflies."
– Stephanie Pace Marshall

We have to choose to transform from the climate *heating* phase to a climate *healing* phase. The climate *heating* "EGO" phase of our civilization was based on the domination paradigm, with man over woman over the animals.

The EGO phase is based on the delusion of separation from Nature. The knowledge that we built up over generations was deployed towards the accumulation of power in fewer and fewer hands, which has led to the *heating* of the planet.

In his book, *The Burning Earth*, Sunil Amrith[1] wrote,

"The most privileged people in the world began to think that the human battle against nature could be won. They believed that natural limits no longer hindered their quest for wealth and power. They believed that instant access to the prehistoric solar energy embedded in fossil fuels made them invulnerable. Their steam engines and lethal weapons conquered the world. In pursuit of freedom, they poisoned rivers, razed hills, made forests disappear, terrorized surviving animals and drove them to the brink of extinction. In pursuit of freedom, they took away the freedom of others. The most powerful

There IS a Planet B

> *people in the world believed, and some still believe, that human beings and other forms of life on Earth are but resources to be exploited, to be moved around at will."*

Now, we have to transform to the climate *healing* phase, where man and woman are serving the animals, restoring their habitats and ensuring their well-being out of love. This is the SEVA phase, which stands for service in Sanskrit.

Judy Carman[2] calls this the transformation from *homo sapiens*, a Latin phrase meaning the "wise hominid" to *homo ahimsa*, the "kind hominid" in the combination of a Latin and Sanskrit word, symbolizing our unity across diverse cultures. In this phase, the knowledge that we built up over generations will be deployed towards the restoration of our planetary home and the rehabilitation of its biodiversity.

Once we cool the planet to within an acceptable range of the temperature that existed in pre-industrial times in the climate *healing* phase, we can transition to the climate *harmonizing* "ECO" phase, where we are equal to all other species.

This is like the caterpillar turning into the chrysalis and then into the butterfly. As Anton Chekhov wrote, *"In Nature, a repulsive caterpillar turns into a beautiful butterfly. But with humans, a beautiful butterfly turns into a repulsive caterpillar."*

However, that perception comes from a short-term view of humanity. Just like the elephant breaking branches off trees turned out to be a benefit to Nature in the long run, we humans are also poised to transform and benefit Nature in the long run. Now we are collectively turning from a "repulsive caterpillar" into the chrysalis of a beautiful butterfly, but for each individual human being, it is a return home to our true childhood selves, when we wouldn't hurt animals unnecessarily.

Rewilding Planet B begins with rewilding our minds, discarding the conditioning from the climate *heating* phase that no longer serves us in the climate *healing* phase of our civilization.

It requires going from a system of *normalized violence* to a system of *normal nonviolence*. Nonviolence is normal for us when we are no longer afraid to be on this earth.

From a frightened prey species, we have become the top predator species on the planet and it is now time to transition to a *caregiver* species.

Rewilding Planet B requires implementing the Seven Core shifts:
1. Speciesism / Colonialism / Racism / Ableism / Patriarchy → **Veganism and Radical Sacredness**
2. Mindset of Scarcity → **Mindset of Abundance**
3. Culture of Consumerism → **Culture of Compassion**
4. Money driven Economy → **Service driven Economy**
5. Diseases and Divisions among humanity → **Health and Unity** among humanity
6. Death and Cruelty towards animals → **Love and Kindness** towards animals
7. Destruction and Pollution of the planet → **Remediation and Regeneration** of the planet

---

[1] This is a book on how humanity has reshaped the planet and the planet has shaped human history over the last 500 years: Amrith, S., *The Burning Earth: A History*, W. W. Norton and Co., Sep 2024. Link https://www.sunilamrithauthor.com/book/the-burning-earth accessed on March 7, 2025.

[2] Judy Carman coined the term "Homo Ahimsa" and made the case that until we liberate animals from human exploitation and violence, we cannot expect to have true freedom and peace for ourselves in 2003. Judy Carman, *Peace to All Beings: Veggie Soup for the Chicken's Soul*, Lantern Press, 2003. Link https://lanternpm.org/book/peace-to-all-beings/ accessed on March 7, 2025.

20. The Core Values Transformation

"Nothing in life is to be feared.
It is only to be understood."
– Marie Curie

To accomplish this greatest transformation in human history, we need to fundamentally transform the core values that drive our civilization.

The first core values transformation is from **deception** to **honesty**. In the climate *heating* phase, deception is prized in the games we play and in the art of persuasion for the marketing of products in the consumer economy.

If a bowler in cricket or a pitcher in baseball is able to deceive opponents into swinging wildly at the wrong line, then they are rewarded handsomely.

If a soccer player is able to deceive opponents into defending in the wrong place, then they are lauded and valued in society.

The climate *heating* phase is about competing and becoming the most powerful species and the most powerful member of that most powerful species. In the climate *healing* phase, honesty will be prized in the games we play and the responsibilities we fulfill as we work collaboratively to heal the climate.

There IS a Planet B

Climate *healing* has a strong engineering component as we strive to address the planetary boundary transgressions and bring the life-support systems of the planet to within the safe zone. In my experience running engineering projects, honesty is the foundation of any successful engineering team.

In a successful engineering team, members should be encouraged to admit when they do something wrong without any repercussions so that mistakes can be corrected. Conversely, team members should be discouraged from hiding their mistake as that would compound the error and ultimately lead to the failure of the project. For our climate *healing* project, failure would mean losing a livable planet and therefore, the stakes are high.

A civilizational framework that encourages deception is the exact opposite of that which can lead to successful climate *healing* endeavors over the next few years.

The second core values transformation is from **domination** to **humility**. In the climate *heating* phase, domination is used in our relationships with each other, with the animals and with Nature in order to gain power over the other as the goal is to become the most powerful species and the most powerful members within that species. In the climate *healing* phase, humility is prized in our relationships with each other, with the animals and with Nature.

Imagine that we were genetically programmed to "serve animals," and we mistook it for the past 10,000 years to mean that we were intended to cook animals and serve them to each other as food. Now we are waking up to the idea that we are here to serve animals and ensure their well-being.

Humility with respect to animals leads to the adoption of Veganism as well as service to animals and nature. It is only when we adopt service as a value that we will be able to routinely contribute more to the planet than we take from the planet and thus begin an upward *healing* spiral in our relationship with Nature.

In the climate *heating* phase, we routinely deal **death** to the animals, cause **diseases** for human beings and **destruction** to the planet,

even though we deceptively claim to strive for **health**, **happiness** and **harmony**. In the climate *healing* phase, we will be truly striving for **health, happiness and harmony** in our relationships with each other, with the animals and with Nature. While death, disease and destruction are not sustainable, **health, happiness and harmony are infinitely sustainable**.

Systems and processes we create in the climate *healing* phase of our civilization on Planet B will be measured against **honesty, humility, health, happiness and harmony**, the 5H values as a litmus test, rejecting any system or process that does not pass this values criterion.

21. The Goals Transformation

"Meaningful reforms are not possible
until goals change."
– Dr. Stephen Kaufman

Foundational Myth on Planet A:
Human well-being necessitates endless economic growth.

Foundational Truth on Planet B:
Human well-being necessitates a sense of purpose larger than
ourselves.

If our goal is to make money and grow the economy as we measure it today, then no matter what we do, we will be stuck in the climate *heating* phase of our civilization and will continue to endure all of the disasters that we see unfolding before us daily.

If our goal is to *heal* the planet, then that must become the overarching goal of our civilization. We must value life over money.

With respect to the UN Sustainable Development Goals[1] (SDGs), we have to drop the redundant SDG #8, *Decent Jobs & Economic Growth* and add SDG #18, *Zero Animal Exploitation* in the climate *healing* phase of our civilization. SDG #18 was proposed by the Beyond Cruelty foundation[2] and if we meet that goal first, it would make

There IS a Planet B

all the other goals easy to meet. The UN states that its overall objective is to obtain peace and prosperity for people and the planet and surely, we can all agree that we cannot achieve peace and prosperity for the planet while continuing to exploit animals unnecessarily.

Economic Growth is a redundant goal because if we meet the other goals — 1) No Poverty, 2) Zero Hunger, 3) Good Health and Well Being, etc. — it is absolutely irrelevant whether the economy is growing or not. Then why is SDG #8 even there, except to deceive people that we are trying to meet all the other goals while we are merely focused on growing the economy?

Economic growth, if it happens, could be an outcome of our efforts as we sincerely try to meet all the other SDGs, but it should not be a goal.

SDG #1 (No Poverty): *End poverty in all its forms everywhere.* This goal can be met with the provision of a Universal Basic Income (UBI) for all humans. Alternately or in addition, if the fundamental needs of healthy food, clean water, adequate clothing and comfortable shelter are met for all humans by the community as part of a set of basic human rights, then all humans would be in a position to restore the ecosystems of the planet and quench the six alarm fire on our planet as part of a set of basic human responsibilities.

SDG #2 (Zero Hunger): *End hunger, achieve food security and improved nutrition and promote sustainable agriculture.* This is one of the easiest goals to meet from a systems engineering perspective.

The world is already feeding over 40 billion land animals[3] with adequate nutrition so that they are able to grow to slaughter weight quickly, diverting half our cropland outputs to feed animals instead of humans. At the moment, we are selectively feeding animals over humans causing food insecurity for 700-800 million humans[4], with

113

around 9 million humans dying each year from hunger or hunger related causes. When we stop exploiting animals unnecessarily, we can eliminate world hunger and ensure adequate nutrition for all 8 billion humans instead.

SDG #3 (Good Health and Well Being): *Ensure healthy lives and promote well-being for all at all ages.* This begins with the availability of healthy nutrition for all. Nature is such a perfect system design that currently we have the means to provide healthy, nutritious whole-foods plant-based vegan meals to every human being on the planet if we so choose. Of course, we also have the means to make available unhealthy, toxic foods to everyone and it is up to us to choose how we use our resource availability.

Besides healthy nutrition, good health and well being also includes the other five pillars of Lifestyle medicine, adequate sleep, regular exercise, refraining from substance abuse, stress management and contact with others. We propose to add a seventh pillar, **A Sense of Purpose**, as essential to meet SDG #3.

SDG #4 (Quality Education): *Ensure inclusive and equitable quality education and promote lifelong learning opportunities for all.* Quality education should now be appropriate for the climate *healing* phase to help humans fulfill their responsibilities as the caregiver species of the planet, instead of teaching them to manage diseases and extract resources from the planet.

SDG #5 (Gender Equality): *Achieve gender equality and empower all women and girls.* Provide gender neutral opportunities for all to contribute towards the well-being of the planet.

SDG #6 (Clean Water and Sanitation): *Ensure availability and sustainable management of water and sanitation for all.* When we are engaged in restoring ecosystems, we would also be restoring the fresh water cycles of the planet making it easier to meet this goal. Besides, when people are consuming primarily whole plant foods, it is easier to implement compostable toilets and create non-toxic soil from human waste.

There IS a Planet B

SDG #7 (Affordable and Clean Energy): *Ensure access to affordable, reliable, sustainable and modern energy for all.* The systems scientist, Donella Meadows[5] estimated that 99% of human activities today are unnecessary and wasteful. Therefore, in the climate *healing* phase, we estimate that the energy needs of humanity would decrease by an order of magnitude as we focus on restoring ecosystems and stabilizing the life-support systems of the planet.

SDG #18 (Zero Animal Exploitation): *Ensure that animals (including humans) are not exploited for human purposes as technology has rendered obsolete the necessity for animal use.* The Beyond Cruelty foundation has documented all the means by which an implementation of SDG #18 makes all the other SDGs so much easier to meet. The inclusion of SDG #18 in place of SDG #8 instils a backbone of integrity to the UN SDGs, while conscripting their universal adoption for ushering in the climate *healing* phase of our civilization.

SDG #9 (Industry, Innovation and Infrastructure): *Build resilient infrastructure, promote inclusive and sustainable industrialization and foster innovation.* When our overall objective is to restore and stabilize the life support systems of the planet, the industrial infrastructure we need for meeting that objective would be far less than that which is needed to grow the consumer economy endlessly.

SDG #10 (Reduced Inequalities): *Reduce inequalities within and among communities.* Climate *healing* as an engineering project has the maximum chance to be successful when there is little perceived inequality within the human team and everyone feels like they are being treated fairly. That is not to say there are no differentiated responsibilities between individuals, but that there are certain aspects related to human dignity where everyone should be treated the same. This is a far cry from what we are experiencing in society in the climate *heating* phase of our civilization today where inequalities are being enhanced.

SDG #11 (Sustainable Cities and Communities): *Make cities and human settlements inclusive, safe, resilient and sustainable.* When the overall objective of the climate *healing* phase is to restore and stabilize the life-support systems of the planet, cities and

communities would need to be designed around meeting that objective. In contrast, in the climate *heating* phase, we gather in cities and communities in order to combine our talents and resources to extract from the planet.

SDG #12 (Responsible Consumption and Production): *Ensure sustainable consumption and production patterns.* When the objective of civilization in the climate *healing* phase is to restore and stabilize the life-support systems of the planet, systems and processes would need to be put in place to encourage humans to give more than we take from the planet. This would result in a societal impetus to embrace minimalism and ensure that this goal is met.

SDG #13 (Climate Action): *Take urgent action to combat climate change and its impacts.* In the perfect system design that is Nature, the urgent action needed to combat climate change and its impacts is also the same urgent action needed to combat biodiversity loss and its impacts, ecosystem collapse and its impacts, fresh water scarcity and its impacts, and so on. It is through the adoption of SDG #18 first and implementing a plant-based human economy.

SDG #14 (Life below Water): *Conserve and restore the oceans, rivers and marine life before it is too late.* At the United Nations and in the climate *heating* phase of our civilization, this goal was about exploiting the oceans, rivers and marine life for human use. The official UN description of this goal reads, *"Conserve and sustainably use the oceans, rivers and marine resources for sustainable development."* However, once we adopt SDG #18, the description of this goal can be altered accordingly.

SDG #15 (Life on Land): *Conserve and restore native ecosystems, combat desertification, halt and reverse land degradation and halt and reverse biodiversity loss.* Once again, in the climate *heating* phase of our civilization, the goal was about exploiting land for human purposes without causing too much more damage. The official UN description reads, *"Protect, restore and promote sustainable use of terrestrial ecosystems, sustainably manage forests, combat desertification, and halt and reverse land degradation and halt biodiversity loss."* Once we adopt SDG #18, this goal is turned outward into what we can

There IS a Planet B

do to help other species instead of being merely about what we can extract from them.

SDG #16 (Peace, Justice and Strong Institutions): *Promote peaceful and inclusive societies with justice for all and build effective, accountable and inclusive institutions at all levels.* In the climate *healing* phase, this goal would include justice for animals, not just for humans.

SDG #17 (Partnership for the Goals): *Strengthen the means of implementation and the global collaboration needed for meeting the goals.* Recognizing that ecosystems restoration is best accomplished with local know-how, this goal is to ensure that the necessary support for local problems is available from global sources, as required.

In this manner, we can redirect the UN SDGs and the momentum that has been built around them towards implementing and accelerating the climate *healing* phase of our civilization on Planet B.

---

[1] The 2030 Agenda for Sustainable Development, adopted by all United Nations Member States in 2015, provides a shared blueprint for peace and prosperity for people and the planet, now and into the future. At its heart are the 17 Sustainable Development Goals (SDGs), which are an urgent call for action by all countries - developed and developing - in a global partnership. They recognize that ending poverty and other deprivations must go hand-in-hand with strategies that improve health and education, reduce inequality, and spur economic growth – all while tackling climate change and working to preserve our oceans and forests. https://sdgs.un.org/goals link accessed on March 7, 2025.

[2] Beyond Cruelty Foundation launched the SDG 18 Campaign in 2018 with the aim of encouraging UN representatives to add SDG 18 – Zero Animal Exploitation to the existing 17 UN Sustainable Development Goals. In turn, motivating nations, organizations, and individuals around the world to adopt policies that seek to reduce and eliminate our reliance on animals. https://sdg18.org/ link accessed on March 7, 2025.

[3] Our World in Data shows 42.8 Billion land animals being farmed in the world at any given point in time: https://ourworldindata.org/grapher/livestock-counts?tab=table Link accessed on March 7, 2025.

[4] According to the WHO/UN FAO, approximately 700-800 million people are hungry and 9 million people die from hunger and hunger-related causes each year. This includes malnutrition, which weakens the immune system and makes people more vulnerable to diseases. The latest WHO/ UN FAO report on *The State of Food Security and Nutrition in the World*. Link https://www.who.int/publications/m/item/the-state-of-food-security-and-nutrition-in-the-world-2024 accessed on March 4, 2025.

[5] Donella Meadows was quoted in the book, Meadows, D., et al., *Limits to Growth: The 30 Year Update*, Chelsea Green, 2004. The link https://www.chelseagreen.com/product/limits-to-growth/ accessed on March 7, 2025.

22. The Self Transformation

"A healthy planet and healthy people are two sides of the same coin."
– Dr. Margaret Chan

When our overall goal is a healthy planet, we must first focus on getting our human family physically, mentally, emotionally and spiritually healthy. This is the oxygen mask rule - put on your own oxygen mask first before helping others. Sick people are not going to care about a sick planet and conversely, a sick planet creates more sick people. Navigating our way out of this vicious cycle is a radical act of rebellion in the current system on Planet A.

The American College of Lifestyle Medicine[1] (ACLM) is *"the medical professional society for physicians and other professionals to practice lifestyle medicine as the foundation of a transformed and sustainable health care system."* The purpose of ACLM is to steer health care practices away from disease management protocols that are widespread on Planet A towards prevention and reversal of chronic diseases through lifestyle change modalities.

The ACLM identified 8 risky behaviors leading to 15 chronic conditions - diabetes, coronary artery disease, hypertension, back pain, obesity, cancer, asthma, arthritis, allergies, sinusitis, depression, congestive heart failure, lung disease (COPD), kidney disease and high cholesterol - responsible for 90% of the health care costs worldwide. The 8 risky behaviors are:

There IS a Planet B

1. **Poor Diet:** According to the late great Lifestyle Medicine pioneer, Dr. John McDougall, poor diet alone is responsible for more than half of the health care costs in the world. The Standard American Diet (SAD) which is now being heavily promoted globally on Planet A, is such a poor diet that it has left in its wake a sick American nation containing the most overweight and obese population in human history, with nearly half of all adults ingesting anti-depressants, anti-anxiety medications or illegal drugs on a regular basis.

2. **Physical Inactivity:** The economic system on Planet A encourages everyone into compulsive behaviors that exercise nothing more than our fingers as we tap our way into the latest rabbit hole on social media, while being fed a steady dose of advertising to promote our adoption of a poor diet. Physical inactivity is second in the list of risky behaviors leading to chronic diseases.

3. **Smoking:** While smoking has now been recognized as a risky behavior for some time and smoking is being officially discouraged in the global North, tobacco companies have successfully pivoted to market their wares in the global South and therefore, globally, smoking ranks third in the list of risky behaviors leading to chronic diseases.

4. **Lack of Health Screening:** The ACLM identifies lack of health screening which results in people unaware of their nutritional deficiencies or the chronic conditions that they are slipping into as fourth in the list of risky behaviors leading to chronic diseases.

5. **Poor Stress Management:** Chronic stress can lead to chronic conditions such as hypertension, heart disease, obesity, Type II diabetes, arthritis and so on.

6. **Insufficient Sleep:** Sleep loss and sleep disorders have a profound impact on human health leading to chronic conditions such as obesity and hypertension.

7. **Poor Standard of Care:** It is now widely documented that inadequate, inaccessible or poor medical care leads to excess deaths due to chronic diseases such as heart disease, stroke and diabetes in certain human populations, specifically in marginalized communities.

8. **Excessive Alcohol Consumption:** The excessive consumption of alcohol and other mind altering substances increases the risk of certain types of cancer, neuropsychiatric conditions and numerous cardiovascular and digestive diseases.

The political and education system on Planet A can be expected to subtly encourage these risky behaviors because on the flip side, they support 90% of the health care economy, which on Planet A, needs to grow endlessly.

On Planet A, in the climate *heating* phase of civilization, we lure our children into poor diet habits with candies strategically placed at eye level in supermarkets and unhealthy foods marketed by clowns handing out free toys at fast food outlets. In addition to children, the adults are also encouraged to consume sickening foods and digital media, exercising nothing more than our fingers, totally detached from the natural world. Mental stress, lack of sleep and substance abuse are the consequences of such a deprived lifestyle.

On Planet A, pharmaceutical companies maximize their revenues when they don't cure our chronic diseases, but maintain them. Likewise, world leaders maximize billionaire donors' wealth when they don't cure the planet's ills, but maintain them. Indeed, if our objective is to grow the health care economy as it is construed today, then we will surely fail to meet our goal of SDG #3, Good Health and Well Being, for humanity.

Climate *healing* communities break this vicious cycle by promoting the exact opposite of the 8 risky behaviors among their members:
1. **From "Poor Diet" to "Food Healers Diet,"** where nutritious, satiating, whole-foods, plant-based vegan meals are made freely available to all in the community as a basic human right.
2. **From "Physical Inactivity" to "Regular Exercise or Yoga Asanas,"** where regular physical activity is encouraged as a matter of routine.
3. **From "Smoking" to "Conscious Breathing or Pranayama,"** where community members re-learn how to breathe deeply and enjoy the life-giving qualities of our earth's atmosphere.

There IS a Planet B

4. **From "Lack of Health Screening" to "Regular Health Routines or Sadhana,"** where community members adopt daily routines that support health.

5. **From "Poor Stress Management" to "Meditation and Awareness or Dhyana,"** where community members learn how to meditate and get in touch with their inner selves.

6. **From "Insufficient Sleep" to "Adequate Rest and Sleep,"** which largely happens when the first five risky behaviors are corrected.

7. **From "Poor Standard of Care" to "A Sense of Purpose,"** as that comes with the consciousness of caring for oneself. Now that we are aware of our responsibilities towards stabilizing and maintaining the life-support systems of the planet, we have acquired a truly altruistic sense of purpose as a species. In alignment with that broader mandate, every community and individual can now find a sense of purpose locally.

8. **From "Excessive Alcohol Consumption" to "Contact with Nature,"** which reduces the feeling of disconnection that causes substance abuse in the first place.

The good news is that we can form such climate *healing* communities starting today and we don't even have to get permissions from anybody. These 8 *healing* habits are closely linked to the Six Pillars of Lifestyle Medicine that ACLM promotes[2]:

1. **Nutrition:** Consuming a fiber-filled, nutrient-dense, antioxidant-rich eating pattern based predominantly on a variety of minimally processed vegetables, fruits, whole grains, legumes, nuts and seeds.

2. **Physical Activity:** Engaging in regular and consistent physical activity.

3. **Stress Management:** Incorporating stress-reducing behaviors may be difficult in modern society but is essential for whole-person health.

4. **Restorative Sleep:** Striving for 7-9 hours of high-quality sleep, allowing the body to reset and recover.

5. **Social Connection:** Strengthening and maintaining relationships and connections with others that bring meaning and purpose to life.

123

6. **Avoidance of Risky Substances:** Reducing or eliminating the consumption of or exposure to any substances that cause harm through toxicity, addiction, physical damage, or adverse side effects.

The 8 *healing* habits includes these six pillars and explicitly adds a seventh pillar, "A Sense of Purpose" and an eighth pillar, "Regular Health Routines," in the context of transforming to the climate *healing* phase of our civilization on Planet B. An altruistic Sense of Purpose towards an objective much larger than ourselves is foundational to human well-being on Planet B.

---

[1] The American College of Lifestyle Medicine (ACLM) is the medical professional society for physicians and other professionals dedicated to clinical and worksite practice of lifestyle medicine as the foundation of a transformed and sustainable health care system. Link https://lifestylemedicine.org/ accessed on March 7, 2025.

[2] The six pillars of Lifestyle Medicine is found here: Link https://lifestylemedicine.org/wp-content/uploads/2023/06/Pillar-Booklet.pdf accessed on March 7, 2025.

23. Busting The Protein Myth

"A myth is a lie that conveys a truth."
– C. S. Lewis

Foundational Myth on Planet A:
Protein is only found in animal foods.

Foundational Truth in Planet B:
All plant foods have all amino acids and have complete protein.

The Protein Myth is the myth that protein is only found in animal foods. This myth is foundational to the economic and political system on Planet A as it drives world hunger and therefore, an abundance of workers at the bottom rung of the economic ladder willing to do work that no one would do otherwise.

A protein is a complex molecule made up of 20-plus amino acids, which the human body can acquire from food sources or by modifying other amino acids. It is the building block of life.

Of the amino acids, 9 are essential for the human body and must be acquired through food sources: histidine, isoleucine, leucine, lysine, methionine, phenylalanine, threonine, tryptophan, and valine. The

There IS a Planet B

human body has the capacity to synthesize the other 11 plus amino acids from these 9 essential amino acids as necessary.

On Planet A, we drill the protein myth into children's minds in our schools, by constantly associating protein with animal foods in science text books. This creates plausible deniability for the authors of children's science text books since they didn't lie about protein being found in animal foods and they can sleep at night. They only forgot to mention that all plant foods have all essential and non-essential amino acids as well[1]. In fact, all plant foods have complete protein.

At the moment, the average American is eating twice as much protein as they need and only one-third of the fiber that they need. Americans and by extension, the whole world, are getting sicker and sicker year by year, as obesity rates continue to soar, mainly due to this fiber deficiency.

Logically, it should be self-evident that plant foods have plenty of protein because elephants, rhinos and gorillas get big and strong by just eating plants. But the vast majority of humans fail to follow this logic when authority figures such as science teachers constantly tell them that protein is found in animal foods.

Even the pedagogy on protein in mainstream nutrition science is skewed to promote the consumption of animal foods. It is based on the premise that the optimum amino acid profile for protein is that found in our muscles, which is closely matched in animal foods, since animals are genetically closer to us than plants. Protein efficacy is typically measured in how fast our muscles grow in response to protein consumption, as if maximizing muscle growth always leads to human well-being.

For a species that has the slowest growth requirement among all mammals, as evidenced by the lowest protein content among all breast milks, this is a dangerously false premise.

Dr. T. Colin Campbell[2] has empirically established that casein, the animal protein found in cow's milk, would be the most relevant carcinogen ever, if it were officially tested among known carcinogens.

Treating animal protein as the ideal is a form of simple-minded logic, like believing the optimum food for an elephant is another elephant and the optimum food for a giraffe is another giraffe. Such simple-minded nutrition science gives science a bad name.

Dr. Walter Willett, the chair of Harvard's Department of Nutrition, said it best[3]: *"To the metabolic systems engaged in protein production and repair, it is immaterial whether amino acids come from animal or plant protein. However, protein is not consumed in isolation. Instead, it is packaged with a host of other nutrients."*

He recommends that we *"pick the best protein packages by emphasizing plant sources of protein rather than animal sources."*

In fact, when we ingest animal foods, it results in our bodies producing higher levels of the hormone Insulin-like Growth Factor-1 (IGF-1).

This hormone stimulates cell division and growth in both healthy and cancer cells and, for this reason, having higher levels of IGF-1 has been consistently associated with increased cancer risk.

Animal proteins also have, in general, higher concentrations of sulfur-containing amino acids, which can induce a subtle state of acidosis when metabolized. One of the mechanisms our bodies use to compensate for this acidosis is leaching calcium from our bones to help neutralize the increased acidity. Over time, this can have a detrimental effect on bone health.

This is thought to be one of the reasons why some studies have found that populations with higher dairy consumption, as well as higher consumption of animal foods in general, also have a higher incidence of bone fractures.

Trimethylamine N-oxide (TMAO) is a compound produced when gut bacteria metabolize certain nutrients found in animal foods such as L-carnitine and choline. High levels of TMAO are associated with increased cholesterol plaques in our blood vessels and therefore, increased risk of cardiovascular diseases and other health issues. Even without all of the other problematic aspects of animal foods, this one issue involving TMAO is, according to the recent president

There IS a Planet B

of the American College of Cardiology Dr. Kim A. Williams, sufficient by itself for people to scrupulously avoid animal foods.

Animal foods contain high levels of phosphorus. And when we consume high amounts of phosphorus, one of the ways our bodies normalize the level of phosphorus is with a hormone called Fibroblast Growth Factor 23 (FGF23).

FGF23 has been found to be harmful to our blood vessels and is associated with heart attacks, sudden death, and heart failure.

Furthermore, there's no need to obsess about getting enough protein from plants either. If we are eating a sensible variety of plant foods, vegetables, fruits, legumes, grains, roots, nuts, and seeds, and we are eating enough calories to be satiated, there is no need to worry about protein adequacy.

The amino acids we need are structurally identical regardless of the source. However, as discussed above, there are serious health implications depending on whether the amino acids are packaged within animal foods or within plant foods.

Plant foods are the real high quality foods that we should be eating for optimal health. This is a part of the protein myth busting that needs to be taught to everyone on Planet B.

---

[1] "There are common misconceptions about whether all plant foods contain all 20 amino acids. It is widely believed among both health professionals and the general population that certain plant foods are entirely devoid of specific amino acids and, thus, that protein adequacy cannot be supported by plant foods alone. In fact, all plant foods contain all 20 dietary amino acids" – Gardner, C., et. al, *Maximizing the intersection of human health and the health of the environment with regard to the amount and type of protein produced and consumed in the United States*, Nutrition Reviews, Volume 77, Issue 4, April 2019. Link https://doi.org/10.1093/nutrit/nuy073 accessed on March 7, 2025.

[2] "In my laboratory research conducted over a quarter century, funded by taxpayer dollars with findings published in the very best journals, we studied this effect in many ways at a most fundamental, cellular and sub- cellular level as much research as for any other chemical deemed to be a carcinogen" – Dr. T. Colin Campbell. Link https://nutritionstudies.org/provocations-casein-carcinogen-really/ accessed on March 8, 2025.

[3] The information in the rest of this chapter on protein is courtesy of Dr. Varsha Shah, Pediatrician and Lifestyle Consultant, who shared with me her course notes on a course for health care professionals at SBKS Medical College, Vadodara, Gujarat, India. It details the known mechanisms through which animal protein and the animal foods that it is packaged in cause chronic diseases.

24. Overcoming the Dairy Deception

*"Deception by an omission of the truth
is as bad as a lie."*
– Jennifer Chiaverini

Foundational Myth on Planet A:
Calcium is only found in dairy foods.

Foundational Truth on Planet B:
If you eat enough calories, you will get enough calcium.

Perhaps the longest running deception in the history of humanity is the notion that cow's milk is good food for humans.

Cow's milk is the lactation secretion of a mammal mother who is five times our size, designed to make her baby grow from calf to cow in 18 months, while humans grow from baby to adult in 18 years. How can that baby calf growth fluid designed to make a body grow 40-60 times faster than the fastest growth ever needed by humans, possibly be good food for humans? Perhaps, it might have been good as medicine once upon a time, but as food?

Yet, dairy is heavily marketed with the help of national and regional governments throughout the world, even in countries where the

There IS a Planet B

vast majority of the human population cannot even digest dairy properly.

What is going on?

For instance, India is the largest producer of dairy in the world and yet, 60 to 80% of Indians are "lactose intolerant". The National Dairy Development Board in India was funded and supported by the European Economic Commission in the 1970s[1], because when people consume dairy, they are paying for the upkeep of the cow on a daily basis and therefore, when the mother cow stops producing milk, she can be turned into cheap beef and leather. She is all paid for by the dairy consumers.

India is the largest producer of dairy in the world and also one of the largest exporters of beef and leather in the world. The European Economic Commission was interested in the cheap beef and leather products and the Indian public was deceived into going along with excessive dairy consumption using religious folklore and other marketing devices, despite the fact that dairy causes them chronic diseases, indigestion, gas and stomach bloating.

The main selling point for dairy foods is that it contains calcium. Once again, the association between calcium and dairy foods is drilled into children's minds through school science textbooks which state that calcium is found in dairy foods. Just as for protein, this creates plausible deniability for the authors of children's science text books since they didn't lie about calcium being found in dairy foods and they can sleep at night. They just forgot to mention that all plant foods also contain calcium, some in lesser and some in greater degree than dairy foods.

When people of authority such as science teachers constantly associate dairy foods with calcium, it leaves the impression that calcium is *only* found in dairy foods. And, the calcium myth is born in impressionable minds.

According to Dr. Faraz Harsini, a molecular biochemist, consuming dairy foods to get calcium is like drinking soda to get potassium[2].

133

Soda may have the nutrient, but it is not the best source of potassium.

Likewise, dairy foods may contain calcium, but they also contain casein, the most relevant carcinogen known to humans as well as a PUS BATH (Pus, Urine, Stools, Blood, Antibiotics, Toxins and Hormones):

Pus: A significant portion of the global dairy herd suffers from mastitis, with reported percentages ranging from 20% to 80% in different communities[3]. Mastitis is an inflammation of the udder caused mainly by the use of machines to milk animals. Machines cannot tell when a teat is inflamed and extract pus along with the milk and when that milk is mixed in with that of other animals, the pus becomes a part of the milk supply.

Different nations have different ranges for the allowable pus cells in commercial milk. The United States Food and Drug Administration (FDA) limits it to 4.4 million pus cells per teaspoon of milk. In Europe, the limit is set at 2.4 million pus cells per teaspoon of milk. In Australia, there is no limit.

Urine: The living conditions for dairy herds are such that they are lying in their own urine and then brought in to milk. As a result, the milk supply contains urine.

Stools: The same living conditions leads to the milk supply containing feces.

Blood: When udders are inflamed as is common in the dairy herd, the milk that is extracted from them contains blood.

Antibiotics: Antibiotics are found in milk through their usage in the treatment of diseased animals as well as from antibiotics being used as feed additives, to prevent diseases and to increase the yield of the herd.

Toxins: The world is swimming in toxins with a fresh load of 220 billion tons injected into the environment each year. These toxins work their way up the food chain in greater and greater

There IS a Planet B

concentrations and can be found in dairy foods at much higher concentrations than in plant foods.

Hormones: Milk is the lactation secretion of a mammal mother who just had a baby. She was most likely also made pregnant again so that she can continue to produce milk year after year. Therefore, naturally, her milk is loaded with hormones designed to grow a baby calf into a great big cow in short order.

As Dr. McDougall put it, *"The result of selling dairy foods to correct a problem that does not exist—calcium deficiency—is that consumers buy foods that actually make them sick."*

Judging by the lactose intolerance percentages around the world, it appears that cow's milk was first consumed by our European ancestors who were probably desperate to stay alive. Today, there are plenty of tropical plant foods available in European supermarkets even in the dead of winter and Europeans are indeed embracing Veganism in droves. Conversely, there are plenty of animal foods available in tropical countries that traditionally mostly ate plant foods and people in the tropics are unwisely embracing animal foods.

It is a test of our intelligence as a species whether we realize that Nature has now given us all the resources we need to create a Vegan world and thrive along with our fellow earthlings on this planet. This requires us to adopt the tropical, abundance-minded ways of living in temperate zones and avoid temperate, scarcity-minded ways of living in tropical zones.

Cows and other bovine mothers have nursed us through the climate *heating* phase of our civilization for millennia. Surely, we are now adult enough to wean ourselves from our mother cow and grow up as a species on Planet B?

[1] The "Operation Flood" program for promoting dairy production and consumption in India was funded with commodity aid from the European Economic Commission. The commodities, such as butter and milk powder, was used as a buffer stock to stabilize market fluctuations and to prime the pump of markets that will later be supplied by domestic production. Link https://documents1.worldbank.org/curated/en/748851468771700148/pdf/308270IN0Milk01ion01see0also0307591.pdf accessed on March 8, 2025.

[2] Source is the Open Letter from Dr. Faraz Harsini to The Shameless Fund. Link https://karlismyunkle.com/2024/06/26/dr-faraz-harsinis-open-letter-addressing-the-violence-and-hypocrisy-of-jonathan-baileys-campaign-with-loewe-jw-anderson-and-the-shameless-fund/ accessed March 8, 2025.

[3] Stanek, P, et al, *A Review on Mastitis in Dairy Cows Research: Current Status and Future Perspectives,* Agriculture 2024, 14(8), 1292; https://doi.org/10.3390/agriculture14081292 link accessed on March 8, 2025.

25. Gaining Independence from Colonialism 2.0

"We need more than a green transition.
We need a new system that puts people and Nature first."
– Helena Gualinga

We all know that during Colonialism 1.0, food scarcity was used to control people in the climate *heating* phase of our civilization, so that people were starved in order to make them do things they did not want to do.

Between 1857 and 1947, during the British Raj, India went through 25 famines, killing over 60 million Indians from hunger or hunger related causes[1]. These engineered famines were used to persuade South Asians to migrate to distant lands as indentured laborers, with the result that many South Asians whose bodies could not store fat easily died off.

Modern South Asians are the descendants of ancestors who could store fat easily and survived these repeated famines. As a result, South Asians comprise 25% of the global population, but 60% of the heart disease patients due to their disposition for fat retention[2]. India is now the largest producer of dairy in the world as well as the diabetes and heart disease capital of the world. This is not a coincidence.

There IS a Planet B

This is **Colonialism 2.0**. Today, hunger is still used to induce people to work on things they don't want to do in the climate *heating* phase of our civilization on Planet A. 700-800 million people around the world are chronically hungry and around 9 million people die of hunger or hunger related causes annually. However, this chronic hunger is not caused by grain tariff policies as it was done during the Colonialism 1.0 era, but through drilling the protein and calcium myths into young, impressionable minds among those who are blessed with food abundance.

Since farmed animals eat 70% of the corn and soybeans grown, world hunger is a choice we make 3 times a day when we eat animal foods. This choice is essentially forced on us because we are lied to in schools about protein and calcium.

Surveys show that 90% of Americans believe that protein is only found in animal foods and calcium is only found in dairy foods[3]. These myths cause otherwise kind, caring humans to preferentially opt for animal foods, causing a lack of adequate nutrition in nations where the grains that they grow are exported to Europe and North America, while their own people starve.

The carnage among the people who are eating animal foods is even more severe than among those who are starving. Over 30 million people die of avoidable chronic diseases mainly due to their excessive consumption of animal foods, almost four times the 9 million who die of hunger or hunger related causes.

As the late Dr. John McDougall put it, *"Animal foods are not our food. Animal foods are full of cholesterol, animal protein and fat, with no starch, dietary fiber or other essential sugars for health. They are infiltrated with big doses of people-poisoning environmental chemicals and loads of infection causing bacteria, parasites and viruses."*

For our wellbeing as well as that of our planet, it is now vitally necessary that we gain independence from this **Colonialism 2.0,** caused by our preferential consumption of animal foods.

Interestingly, animal foods have been the centerpiece of festivals and celebrations in human societies during the climate *heating* phase of our civilization. This may be because our ancestors did not know about Vitamin B_{12}, but knew empirically that consuming this critical vitamin is necessary for human wellbeing. Vitamin B_{12} was discovered in the 1950s and supplements were commercially available from the 1960s onward.

Vitamin B_{12} is stored in the liver and it is used for vital functions such as the production of red blood cells in our bodies. In nature, Vitamin B_{12} is produced by soil bacteria as well as microorganisms found in the small intestines of animals. Humans are the only species that washes their food before eating and therefore, we don't get much Vitamin B_{12} in the plant foods we consume.

Animals raised on healthy, vibrant soils would not be deficient in Vitamin B_{12} and therefore, the ritualized consumption of animal foods during festivals and celebrations by everyone, rich and poor, may have been sufficient to replenish Vitamin B_{12} stores among the general population in olden times.

Today, the vast majority of people deficient in Vitamin B_{12} are not Vegans, but those who consume animal foods. Due to the poor health of our soils nowadays, farmed animals are also typically deficient in Vitamin B_{12} and therefore, most Vitamin B_{12} supplements are fed to farmed animals, not humans.

Vegans have cut out the middle animal and ensure their adequate intake of Vitamin B_{12} either through fortified foods or a regular supplement.

---

[1] These statistics were sourced from the Guardian's reporting: https://www.youtube.com/watch?v=z8Qv7zZBxq8 link accessed on March 8, 2025.

There IS a Planet B

---

[2] This statistic is found in the cover story of the American College of Cardiology from May, 2019. Link https://www.acc.org/latest-in-cardiology/articles/2019/05/07/12/42/cover-story-south-asians-and-cardiovascular-disease-the-hidden-threat accessed on March 8, 2025.

[3] The latest survey was conducted by the Physicians Committee for Responsible Medicine in Jan 2025: Link https://www.pcrm.org/news/news-releases/nearly-90-us-adults-believe-inaccurately-its-important-eat-animal-products-get was accessed on March 8, 2025.

26. The Oxford Union Debate

*"In all debates, let truth by thy aim,
not victory, or an unjust interest."*
– William Penn

We had a debate on the topic of whether the world should Go Vegan at the Oxford Union on November 30, 2023. Here's the text of my speech during the debate:

"Ladies and Gentlemen, it is my privilege to speak at a venue where, 200 years ago, you began rebelling against false orthodoxy[1]. Today, I want to rebel with you against false orthodoxy by speaking on bovine matters.

I mean, of course, cows. Yes, there is a Cow in the Room and not everyone can see it. I hope that by the end of the debate eyes will be opened.

The orthodoxy, the herd opinion if you like, is that animal agriculture has little to do with climate change. I believe that is very wrong.

There IS a Planet B

I believe that based on data. I am an environmentalist by occupation, but a systems engineer by profession. I invented the protocol for transforming early analog internet connections to more robust digital connections, while accelerating their speed ten-fold. Still today, any data accessed on the internet likely passed through a device implementing this protocol.

I plead that this House rebel once again by voting for the proposition, "This House Would Go Vegan."

Veganism is defined[2] as a *"philosophy and way of living that seeks to exclude, as far as is possible and practicable, all forms of exploitation of animals for food, clothing or any other purpose."*

The proposition asks that this House Go Vegan, not Be Vegan, implying that this is a journey, not a destination. I highly recommend this journey on ethical, health and environmental grounds.

I will now focus on the environmental reasons to Go Vegan.

It is undeniable that human civilization has adversely impacted life-support systems on the planet. Scientists have identified nine planetary boundaries that we must stay within for the sustainability of life on earth. At the moment, we have transgressed six of them[3] and any one of these transgressions could end life as we know it.

The good news is that when we Go Vegan, we help resolve all six of them. That's the power we have as individuals to reverse our existential crisis.

Animal agriculture is the leading cause of ecological destruction because it uses 37% of the ice-free land area of the planet just to graze animals[3], while bottom trawling an area of the ocean floor the size of South America every year for industrial fishing[4].

Animal agriculture is the only major activity in which we destroy forests and replace them, not with other trees for timber or paper, but with grass, which drastically reduces the diversity of life that the land can support.

Animal agriculture is the primary reason why humans have reduced the number of trees on the planet by half, from 6 trillion to 3 trillion, over the past 10,000 years[5]. Restoring those 3 trillion trees can draw down enough carbon to completely reverse climate change.

Animal agriculture is grossly inefficient because animals must eat 39 pounds of plants to produce one pound of human food, on average[6], a burden which the world can no longer afford.

By going Vegan, we can give nearly 40% of the ice-free land area of the planet, as well as the entire ocean, back to nature.

When we restore the native ecosystems on that land, we can grow most of the 3 trillion trees that we cut down over the past 10,000 years. This helps resolve all six planetary-boundary transgressions.

The least violated transgression is fresh-water change. Rewilding the land that is currently used for grazing animals will restore the fresh-water cycles of the planet.

The next is land-system change. Going vegan will allow us to return nearly 40% of the land area of the planet back to nature, resolving this planetary-boundary transgression.

The next worst transgression is climate change, which can be resolved when the excess carbon in the atmosphere is absorbed in the trees and soil that we can restore to the ecosystems of the planet.

The next is chemical pollution, which would be safely stored away in regenerating forests when we Go Vegan.

There IS a Planet B

Eating animal foods currently delivers concentrated doses of this chemical pollution into our bodies through bioaccumulation. Therefore, going Vegan addresses chemical pollution for both the Earth and ourselves.

The next worst transgression is nitrogen and phosphorus loading, mainly through our overuse of synthetic fertilizers for crops. Since over half the crop outputs are fed to farmed animals, going Vegan will resolve this transgression as well.

All of these transgressions impact wildlife, and biodiversity loss is the worst of the six planetary-boundary transgressions.

By restoring habitats for wild animals and allowing them to live freely in the ocean, we will resolve this transgression as well. If instead, we let wild animals die, we die.

There are two explanations for perhaps the gravest threat ever posed to civilization — and all life on earth: — the imminent danger of runaway climate change[7].

One explanation–the one we hear about all the time from our leading climate spokespeople–is the burning of fossil fuels. It is certainly true that the burning of fossil fuels contributes greenhouse gasses to the atmosphere, thereby warming the planet.

But the other explanation — what I call the Cow in the Room — is rarely addressed: the human folly of exploiting animals that is animal agriculture.

When these two sources of greenhouse gasses are compared in the media, fossil-fuel burning is always emphasized, and is almost always assigned the greater responsibility for warming the planet.

But the opposite is true. When you factor in the potential carbon absorption of the forest land cleared for animal agriculture, you find, with any honest accounting — as I

published in a peer-reviewed paper — that animal agriculture is responsible for at least 87% of greenhouse gasses on an annual basis[8].

When I made that calculation, I did not include the respiration of farmed animals. I did not include the bottom trawling of the oceans by industrial fishing. I did not include the carbon released by pasture-maintenance fires set annually on grazing lands around the world. I did not include the loss of phytoplankton populations and sea forests due to industrial fishing.

I did not include those factors mainly because they haven't been reliably assessed due to a futile attempt by the orthodoxy to hide the Cow in the Room. But it seems clear to me that if we could estimate these factors and include them in the calculation, we would find that animal agriculture is responsible for — wait for it — well over 100% of greenhouse gas emissions into the atmosphere.

Now that sounds unbelievable. How could it possibly be responsible for more than 100%? Because the evidence points to the possibility that the earth will start cooling in a Vegan world even if we continue to conduct all our other activities as we do today.

The cessation of animal agriculture will result in healthy oceans, healthy forests and healthy soils, and if we want to reverse climate change, then we must adopt a strategy that can draw down greenhouse gasses from the atmosphere. Healthy oceans and sea forests can do that. Healthy soils and trees can do that. Solar panels and electric cars cannot.

Now, I am not a supporter of the fossil-fuel industry–far from it. It is my engineering assessment that we must wean ourselves off fossil fuels gradually. But we burn fossil fuels to heat and cool our homes, to transport ourselves, to manufacture goods, to ship goods. These are all social goods.

There IS a Planet B

What social good comes from animal agriculture? Nothing. Only obesity, heart disease, diabetes, cancer, biodiversity destruction, soil depletion, fouling of our waterways, antibiotic resistance, dangerous and dehumanizing work, animal cruelty, climate catastrophe, world hunger and let's not forget pandemics. Indeed, there is nothing that will not improve when we end the cruelty and folly of exploiting animals.

I have just given you the intellectual reasons to Go Vegan, but lasting change comes not from the head, but from the heart. In that regard, I have made a pinky promise to our granddaughter, Kimaya, that the world will go largely Vegan by 2026, which is the year we will have killed almost all the wild animals if we don't change course.

Ladies and Gentlemen, I am confident that true to your 200-year tradition of rebelling against false orthodoxy, this House will once again break away from the herd, see the Cow in the Room and vote for the proposition to help our generation keep this sacred promise for all the children of the world.

Thank you for your consideration, from the bottom of my heart!

Our side won the debate. Dozens of elite universities in the world have also called for their universities to Go Vegan, with more being added every month. Therefore, the youth of the world know that the future is Vegan.

When we know that the future is Vegan, why not embrace that future right away instead of procrastinating and letting the planetary fire get worse, day by day?

[1] https://www.youtube.com/watch?v=6HJetz7S0m4 is a video that details the rebellious origins of the Oxford Union. Link accessed on March 8, 2025.

[2] https://www.vegansociety.com/about-us/history describes the history of the Vegan Society. Link accessed on March 8, 2025.

[3] Arneth, A., et. al, IPCC Special Report on Climate Change, Desertification, Land Degradation, Sustainable Land Management, Food Security, and Greenhouse gas fluxes in Terrestrial Ecosystems, UN Intergovernmental Panel on Climate Change, Aug 2019. Link: https://www.ipcc.ch/site/assets/uploads/2019/08/Fullreport-1.pdf accessed on March 7, 2025.

[4] L. Watling, E. A. Norse, "Disturbance of the seabed by mobile fishing gear: A comparison to forest clearcutting," Conservation Biology, vol 12, no 6, Dec 1998, pp. 1180-1197. Link https://marine-conservation.org/archive/mcbi/Watling_&_Norse_1998.pdf accessed on March 8, 2025.

[5] Crowther, T.W., et. al, *Mapping Tree Density at a Global Scale*, Nature 525, Sep 2015, pp. 201-205. Link https://www.nature.com/articles/nature14967 accessed on March 7, 2025.

[6] Smith, P., et. al, *Agriculture, Forestry and Other Land Use (AFOLU)*. In: Climate Change 2014: Mitigation of Climate Change. Contribution of Working Group III to the Fifth Assessment Report of the Intergovernmental Panel on Climate Change, Cambridge University Press, Cambridge, United Kingdom and New York, NY, USA, 2014. Link https://www.ipcc.ch/site/assets/uploads/2018/02/ipcc_wg3_ar5_chapter11.pdf accessed on March 7, 2025.

[7] Armstrong-McKay, D. I., et al., *Exceeding 1.5°C Global Warming Could Trigger Multiple Tipping Points,* Science, vol. 377, no. 6611, Sep. 2022. Link https://www.science.org/doi/10.1126/science.abn7950 accessed on March 7, 2025.

There IS a Planet B

---

[8] Rao, S., *Animal Agriculture is the Leading Cause of Climate Change: A Position Paper*, Journal of Ecological Society, vol 32-33, 2021. The paper is also hosted on the Climate Healers website here: https://climatehealers.org/the-science/animal-agriculture-position-paper/ Link accessed on March 7, 2025.

27. Codes for a Healthy Earth

"In every deliberation, we must consider the impact
of our decisions on the next seven generations."
– Iroquois Wisdom

We begin with adopting *Codes for a Healthy Earth*[1] for implementing the transformation into the climate *healing* phase of our civilization. *Codes for a Healthy Earth* is a unifying, whole-system healing framework that is open to ongoing evolution and refinement. The vision for "the Codes" was inspired by the work of the Global White Lion Protection Trust[2]. The Codes were initially drafted by Dr. Shelley Ostroff and Yan Golding from Together in Creation[3], and has evolved through a consultative and collaborative process with a growing global network of partners from over 30 countries on all continents. The framework can be freely adopted — in its integrity — by any group or movement as a collective compass for coordinated community-led action towards the healing and regeneration of the planet and all its inhabitants.

Codes for a Healthy Earth is inspired by and builds on numerous pioneering declarations, charters, manifestos and guiding principles including The Earth Charter[4], The Universal Declaration of Human Rights[5], The Universal Declaration of the Rights of Mother Nature[6], The Universal Declaration on the Rights of Indigenous Peoples[7] and

There IS a Planet B

others. Together, these documents offer a wealth of wisdom, shared values and aspirations, principles, policies and solutions that could — if implemented — ensure the health and vitality of the planet and all its inhabitants for generations to come.

Codes for a Healthy Earth launched globally on International Peace Day, September 21, 2019. The "Codes" are founded on the premise that as long as there is oppression anywhere within the Community of Life, true peace cannot exist. To achieve genuine and lasting World Peace, it is essential that we cultivate Peace with All of Life.

The *Codes for a Healthy Earth* for cultivating peace with all of life are reproduced in their entirety below:

PREAMBLE

The urgent and complex global challenges we face will not be resolved from within the same systems that created them. Today, people of all cultures and ages are rising up around the world to demand a fundamental transformation of how we organize ourselves as a species.

Hundreds of millions of people and millions of groups are working on countless regenerative and compassionate solutions. Throughout this vast and diverse global movement there is a growing recognition that we already have the knowledge, skills, ideas, technologies and resources, as well as the wise, service-based leadership, to effectively address all our escalating crises. Our primary challenge is to align and organize effectively for whole-system healing and transformation.

To ensure an optimal integration of our global abundance of wisdom and solutions, it is essential that We, the People, reclaim our individual and collective authority and responsibility to align ourselves and our social systems — governance, law, economics,

media, education, etc. — with the principles of Life and evolving human consciousness.

In light of this recognition, we, People of Earth, are uniting around a whole-system healing framework that effectively supports community-led self-organization at the local and global levels to realize our shared needs and aspirations for a Healthy Earth.

DECLARATION

We, People of Earth, unite in love and concern for our planetary home and all its inhabitants. We come together as one humanity, across national, cultural and ideological boundaries, to restore the well-being of all Life on Earth.

We recognize that our personal, collective and planetary health are all interconnected and interdependent. For humanity to thrive, the entire planetary ecosystem must thrive.

We affirm that the only legitimate purpose of governance is to protect and cultivate the health and vitality of the planet and all its inhabitants for generations to come.

THEREFORE, we, People of Earth, pledge to self-organize for immediate and resilient global action. We commit to working together with aligned communities, organizations, institutions and networks to:
1. Transform our social systems to effectively serve the healing, integrity and well-being of all Life.
2. Restore the health and diversity of the biosphere.
3. Ensure all humans and animals can meet their Core Needs through guaranteed access to:
   i. living soil
   ii. healthy water
   iii. vitalizing food
   iv. fresh clean air
   v. physical and emotional safety
   vi. comfortable shelter
   vii. the cultural and ecological conditions and resources needed for all to realize their unique potential — in

There IS a Planet B

    mutual enrichment with their communities and ecosystems.

4. Cultivate a shared, whole-system understanding of our collective challenges, existing solutions, and optimal healing pathways forward.

5. Co-create a globally coordinated transition to a shared culture of peace, kindness, harmony in diversity, wisdom, integrity, accountability, collaboration, regeneration and reverence for All of Life.

GUIDING PRINCIPLES

We commit to organizing around the following principles as the foundations essential for the radical whole-system healing strategy that is urgently required at this time:

Nonviolence and Stewardship

- Aspire to do no harm, and to align our thoughts, words and actions with that which is nourishing for the entire Community of Life.
- Protect, honor and restore the Foundations of Life – Earth, Water, Fire, Air, Climate, Biodiversity, and the Web of Life.
- Protect and honor Indigenous peoples and their lands and integrate their counsel, wisdom and knowledge systems in all sectors of human affairs.
- Ensure all decisions are based on the analysis of consequences to the health of the whole for the next seven generations, and that human and non-human stakeholders are represented and taken into account.

Harmony in Diversity

- Honor and heed the voices of marginalized groups in co-creating the way forward.
- Ensure the full protection and equality of girls and women in all social sectors.
- Develop our understanding about the essential and vitalizing nature of human and ecological diversity.
- Cultivate cultures of inclusivity that celebrate our life-enriching differences.

- Acknowledge the deep layers of individual, cultural and environmental trauma, and work to implement holistic healing and rehabilitation strategies that prioritize the most vulnerable communities, species and ecosystems.

Economy and Law
- Give legitimacy to those laws that protect and regenerate Life and delegitimize all laws that permit harm to Life.
- Invest only in that which protects and regenerates Life, and disinvest from any and all enterprises that harm or could harm Life.
- Ensure all corporations and enterprises harming Life will be supported to either transform or dissolve.
- Create the optimal conditions for the core needs of all humans and animals to be met as we transition to healthier economies and enterprises.

Education, Learning and Media
- Inform people everywhere about the global and local challenges, their systemic causes, and the wealth of existing solutions, initiatives and networks from all sectors dedicated to personal, collective and planetary healing and transformation.
- Identify, develop and scale holistic healing and regenerative learning opportunities for people of all ages, learning styles, interests and abilities, to support the flourishing of the individual, the community, and the whole.
- Cultivate the wisdom, art and science of systems-thinking and whole-system health and healing.
- Evolve our language, communication skills, creative expression and cultural narratives to cultivate an informed, empathic, compassionate and vibrant human society.

Transitioning to New Social Systems
- Ensure that people in positions of leadership are skilled in applying the principles of whole-system health and healing, and fully embody the values of integrity, courage, compassion, wisdom, commitment, service, expertise and dedication to the good of the whole.

There IS a Planet B

- Initiate evolving online platforms to learn, gather, refine and disseminate the best existing innovative laws, policies and practices for whole-system health and healing in all sectors, and to design pathways for their swift worldwide adoption and implementation according to local needs and conditions.
- Align technology and scientific advancement to serve whole-system health and healing.
- Cultivate self-organizing community-led 'wisdom and expertise circles' to advise on the wisest policies and most effective solutions, while giving voice to the needs, resources and interdependencies of all human and non-human stakeholders.
- Design and implement a seamless transition to new forms of local and global governance dedicated to cultivating the most efficient and vitalizing flow of energy (resources, skills, information, ideas, etc.) through the entire planetary ecosystem.

TOWARDS A THRIVING WORLD FOR ALL

We, People of Earth, affirm that a worldwide adoption of the principles outlined in these Codes has the capacity to solve multiple crises simultaneously and create a culture of Peace with All of Life.

We support the translation of these principles into an evolving set of recommended laws and policies for all national and local governments, corporations, and community-led initiatives. We honor all the wisdom traditions and call on them now to bring their unique and shared understanding in service to humanity and all of Life. Aligning around a comprehensive framework for whole-system health enables people across the world to self-organize effectively towards manifesting a shared vision of a thriving planetary ecosystem where:

- The Web of Life is restored to its natural balance, with clean air, healthy soil, and easily accessible pure water and vitalizing food for all living beings.
- All humans and animals receive precisely what they need to realize their unique potential, in mutual enrichment with the whole.

- Global society organizes according to new forms of distributed governance that puts whole-system health as its core purpose and organizing principle.
- Humanity works together as a compassionate life-enriching species, contributing to ever-evolving harmony in diversity, and cultivating the conditions for All of Life to flourish.

TOGETHER WE RISE FOR ALL OF LIFE

You are invited to consider endorsing and sharing *Codes for a Healthy Earth* as a universal and evolving social contract to be voluntarily adopted by all Peoples of Earth who commit to uniting and working together, in mutual responsibility and accountability, for rapid social and ecological healing and regeneration on Planet B.

---

[1] As one of the 300 plus organizations to have endorsed the **Codes for a Healthy Earth** since it was first offered by Dr. Shelley Ostroff in September 2019, Climate Healers has been promoting the Codes as the foundation for governance in the climate **healing** phase of our civilization. Link https://www.codes.earth/thecodes was accessed on March 8, 2025.

[2] The Global White Lion Protection Trust was founded by Linda Tucker, who says, "Today, all conservation issues are global issues. If Brazil loses its rainforests, the world has lost its lungs. If South Africa loses its white lions, the world has lost its heart." Please find more here: https://whitelions.org link accessed on March 9, 2025.

[3] Together in Creation, initiating and supporting pathways for transitioning to a new form of local and global governance that is dedicated to protecting and cultivating the health and vitality of the planet and all its inhabitants for future generations, is the brainchild of Dr. Shelley Ostroff and Yan Golding. Please find more here: https://www.togetherincreation.org/ link accessed on March 9, 2025.

There IS a Planet B

---

[4] The Earth Charter is a document with sixteen principles that drive a global movement towards a more just, sustainable and peaceful world. For more, https://earthcharter.org/, link accessed on March 9, 2025.

[5] The UDHR is a milestone document in the history of human rights, proclaimed by the UN General Assembly in 1948 as a common standard for all peoples and nations. It sets out fundamental human rights to be universally protected and has inspired more than seventy human rights treaties. Please find more here: https://www.un.org/en/about-us/universal-declaration-of-human-rights link accessed on March 9, 2025.

[6] The Universal Declaration for the Rights of Mother Earth was adopted at the World People's Conference on Climate Change and the Rights of Mother Earth held in Colombia, Bolivia on he 22nd of April 2010. This is one of the main sources of the Rights of Nature movement. For more, please see https://www.rightsofnaturetribunal.org/ link accessed on March 9, 2025.

[7] The UN Declaration on the Rights of Indigenous Peoples establishes a universal framework of minimum standards for the survival, dignity, wellbeing and rights of the world's indigenous peoples, adopted by the UN General Assembly on Sep 13, 2007. Only the United States and the Vatican have failed to ratify this UN DRIP. For more, please see https://www.un.org/development/desa/indigenouspeoples/wp-content/uploads/sites/19/2018/11/UNDRIP_E_web.pdf , link accessed on March 9, 2025.

28. The Civilizational Transformation

*"There is nothing as powerful
as an idea whose time has
come."*
– Victor Hugo

We know that most people with chronic conditions as well as the entire planet can be healed if we all start doing the right things — eat right, exercise regularly, sleep soundly, implement the *Codes for a Healthy Earth* and generally lead a life where we routinely give more to the planet than we take from the planet. The difficulty is not in knowing what to do, but in actually getting people to do it, especially when they are being constantly bombarded with contrary messages from vested interests on Planet A.

Just as with individual health, achieving real planetary health takes time, effort and discipline, while hiding symptoms with band-aids may seem like the easier path to take. However, the more we succumb to the short term panacea of the band-aid approaches, the harder it gets to truly solve the underlying problems.

The path forward to a bright future is narrow and hard, but our best chance for achieving it is when everyone feels like they belong in Nature and in the same team called *homo ahimsa*. Besides, nothing worthwhile of this magnitude was ever accomplished without a

There IS a Planet B

whole lot of people pitching in with their expertise towards the common cause.

A Global Infrastructure Upgrade Project

The transformation from a climate *heating* civilization to a climate *healing* civilization can be viewed as a global infrastructure upgrade project. At Climate Healers, we believe that it requires upgrading the educational infrastructure, the healthcare infrastructure, the governance infrastructure, the economic infrastructure, the ecological service infrastructure and the spiritual infrastructure of our civilization, in addition to the energy infrastructure. We believe that this whole-systems infrastructure upgrade project is a social necessity in order to free humans and animals from untold suffering and likely near-term extinction.

We are not alone in this belief. In its citation designating me as a "Climate Hero," the Guardian Newspaper recognized me as a **"foremost voice on the green transition and the true scale of societal change required to save the planet."** Therefore, at least some people at the Guardian Newspaper believe this as well.

Let's consider these infrastructure upgrades one by one.

Sustainability would be built into the educational curriculum in every discipline in a climate *healing* civilization instead of being relegated into separate Centers for Sustainability at a few elite institutions, while the rest of the departments teach archaic, unsustainable practices.

Preventive care and lifestyle medicine would be the healthcare norm in a climate *healing* civilization instead of disease management and symptom alleviation.

The governance of commons based on the *Codes for a Healthy Earth* would be the upgrade from the governance of private property rights and the nation states model, a legacy of the colonial era.

159

On the economic front, *vegan donut economics* would be the upgrade from capitalism with its unsustainable, infinite growth economics and the resultant ever-worsening planetary boundary transgressions.

Humans would be engaged in ecosystems restoration in a climate *healing* civilization instead of treating nature as a basket of resources, ripe for exploitation.

Vegan spirituality that considers humans as an integral part of Nature would be the upgrade from human exceptionalism and separation from Nature as prevalent today, rounding out the whole systems, consciousness upgrade.

Finally, the fossil fuel infrastructure would be gradually phased out in a climate *healing* civilization, upgraded with energy efficiency and renewable energy sources.

As with any infrastructure upgrade project, we can expect inconveniences in each of these realms as the project is executed. There is reason to believe the public will put up with such inconveniences for the promise of a brighter future, provided they see the project being executed with competence and integrity.

Unfortunately, that's a far cry from what has been going on at the United Nations (UN) on Planet A.

The current approach to the necessary infrastructure upgrade of our civilization has been rife with political interference and systemic corruption as entrenched interests attempt to sell us the fairy tale that the only upgrade we need is in the energy infrastructure. Even the science has been tampered with line-by-line editing of the UN scientific reports by political operatives affiliated with member nations at the UN.

As a systems engineer, I do not expect an *Inter-governmental* Panel on Climate Change (IPCC) to report accurately on the science. It is just systems engineering common sense that political operatives affiliated with entrenched interests should not be allowed to edit

There IS a Planet B

scientific reports if we plan to execute successful engineering projects with the contents of the reports.

Is it any wonder that we have witnessed 30 years of frustrating incompetence on our ecological crisis at the UN? As talk of a climate apocalypse heats up while the UN Conferences Of the Parties (COPs) continue to kick the can down the road and multiple wars rage simultaneously around the world even as ecological planetary boundary transgressions worsen alarmingly year after year, it is instructive to consider the tale of two metro infrastructure projects in India as an object lesson for our time.

The Kolkata metro[1] was commissioned in 1972 by Prime Minister Indira Gandhi and it took 12 years to get it up and running. Fifty years later, it had two operational lines, one north-south and another east-west, which don't yet intersect with each other. From the start, the Kolkata Metro project was rife with political interference, systemic corruption, cost overruns and delays.

In contrast, the Delhi metro[2] was commissioned in 1996 by Prime Minister Deve Gowda and it took just 6 years to get it up and running. The project was executed under budget and ahead of schedule. Twenty five years later, it rivaled the London Underground in extent and carried 50% more passengers than its London counterpart, 6 million per day. Today, an estimated 99.97% of the trains on the Delhi metro arrive within one minute of scheduled time and at peak hours, the trains run every 2.5 minutes.

How did Delhi metro do it?

The Delhi metro project absorbed the lessons learned from the Kolkata metro project and then broke the mold.

When Elattuvalapil Sreedharan was approached by the Lieutenant Governor of Delhi to head the Delhi metro project, he knew that political interference would be detrimental to the success of the project. He stipulated that if he was to lead Delhi metro, he should be given complete freedom to run the organization. There shouldn't be any interference from politicians or bureaucrats and that he

should be able to choose his own people. The Lieutenant Governor agreed without any hesitation because important politicians were routinely experiencing horrendous delays in traffic jams in Delhi.

Sreedharan hired motivated staff, solely on merit. Deadlines and budgets were realistic and achievable. Once a decision was made, it was final. If anything went wrong, people looked for solutions, not scapegoats.

There was zero tolerance for corruption. Anyone caught was out immediately. Emphasis was placed on speed, efficiency and getting it right the first time. Safety was paramount on project sites.

Sreedharan viewed the Delhi metro project as a social necessity to free people from time losses, inconvenience, pollution and congestion. From the early stages itself, he used the media to project a positive image of the Delhi metro to the public. As the project repeatedly met deadlines and adhered to stringent safety measures, peoples' hearts were won over.

In short, the Delhi metro was an infrastructure upgrade project executed with competence, engineering integrity and public support.

Surely, we have the wherewithal and the personnel to implement the necessary global infrastructure upgrade from the climate *heating* phase to the climate *healing* phase of our civilization with the same level of competence, engineering integrity and public support as the Delhi metro?

What will it take for global political leaders to emulate former Prime Minister Deve Gowda of India and empower such an independent infrastructure upgrade project for our civilization, free from political interference? What are they waiting for?

What are we waiting for?

The Seven Strategic Actions

We can all plunge in and assist with the seven strategic actions that are already being implemented around the world right now:

There IS a Planet B

1. Education, education, education:

This means telling this new story of Planet B, shining the light on the greatest transformation in human history, exposing the four foundational myths underlying the climate *heating* phase and replacing them with the four foundational truths of the climate *healing* phase of our civilization.

The good news is that education on the protein myth, the calcium myth and the endless economic growth myth, the first three foundational myths of the climate *heating* phase, has been going on for decades, if not for centuries. In the modern, hyper-connected world, there are literally millions of amazing individuals, organizations and documentaries dedicated to helping people overcome these myths.

The late Dr. John McDougall, a pioneer in this area, had been speaking out about the protein myth and the calcium myth while promoting a starch-based, whole-foods vegan diet since the 1970s to reverse and prevent chronic diseases.

Born in 1947, Dr. McDougall suffered a stroke at the age of 18, which left him with a lifelong disability and a hunger to understand the root cause of diseases so that they can be truly cured. He was a passionate truth seeker, who unearthed the four deadly dietary deceptions: the Protein deception, the Calcium deception, the Omega-3 deception and the Carb deception, which he highlighted in his McDougall Research and Education Foundation website[3]. His collection of ten favorite one-liners is a classic piece that would be helpful for all to read and understand[4].

Dr. McDougall healed tens of thousands of people directly and millions indirectly through his books, videos and most importantly, the McDougall program, now managed by his daughter, Heather McDougall. After going through the program, I feel that the McDougall program should be taught in every high school in the world. It is criminally negligent of our education system that children are not taught such basic health-promoting information,

163

especially the information on how to protect themselves from the pleasure trap in our world of food abundance.

The *pleasure trap*[5] is why we do things that are not in our long-term best interests. We naturally seek out foods that give us the most calories and pleasure for the least amount of effort expended in consuming them. As Dr. Lisle points out, the calorific breakdown of the Standard American Diet (SAD) is 51% from oil and refined flour, 42% from animal foods, 3% from potatoes in the form of chips and french fries and 4% from whole fruits and vegetables. Therefore, 95% of the food we eat in the SAD is steering us towards illness not wellness.

Dr. Doug Lisle and Dr. Alan Goldhamer have been helping people overcome this pleasure trap and they have been curing thousands of patients of their chronic illnesses at True North Health center in Santa Rosa, CA. They define 5 phases in the process of people falling into the pleasure trap and coming out of it:

Phase 1: The normal phase where we get normal pleasure from eating starch-based whole plant-foods, the foods that we are biologically designed to eat.

Phase 2: The phase where we get enhanced pleasure from eating drug-like ultra-processed foods and animal foods with increased calorie density, extra sugar, fat and salt.

Phase 3: Once we get used to these drug-like animal foods and ultra-processed foods, we no longer get the same enhanced pleasure, but just normal pleasure from eating them. These become the default food for our palate.

Phase 4: In the fourth phase, we get lowered pleasure from eating whole plant-foods, the foods that we are supposed to be eating and that are good for us. These foods no longer taste good to us as our palate craves for the sugar, fat and salt.

Phase 5: After 8 to 10 weeks of sticking to eating starch-based whole plant-foods, these foods now provide us with normal

There IS a Planet B

pleasure as our taste buds adjust and we overcome our addiction to the drug-like foods.

The best way to overcome the pleasure trap is to first know that there is one. Then, either employ sheer grit to endure a starch-based, whole plant foods diet for 8-10 weeks until our taste buds adjust or find a chef like Marlene Watson-Tara[6] or Dr. Sheil Shukla[7], for example, and/or copy published recipes from such culinary geniuses to create healthy gourmet meals that would help us leap-frog over the difficult **Phase 4** of the pleasure trap. In the latter case, we could involve our whole families and turn this transition into an adventure.

In either case, overcoming the food pleasure trap is not as hard as overcoming addictions to alcohol, drugs or smoking, when we set an intention to overcome it.

Experts in the field of ecological economics and behavioral economics have been educating people about the human wellbeing myth, proving that it does not depend upon endless economic growth. There is a drug called approval or social media likes or followers that we are taught to rely upon for our self-worth, which drives this endless economic growth paradigm. In this regard, it is my privilege to acknowledge the great James Baldwin who taught me in a video interview that what other people think of me is **none of my business.**

Such a mindset is physically, mentally and emotionally liberating and it should form the foundation of any educational training program for climate *healers*.

The only foundational myth that does not have a large number of individuals and organizations working on educating people is the myth about fossil fuels being the leading cause of climate change. This is the newest myth that the system in the climate *heating* phase cooked up on Planet A in order to prop up the other three foundational myths.

During the 1990s, the people in power must have concluded that the nature and climate crisis had the potential to upend the entire

165

climate *heating* phase on Planet A, since animal agriculture is the leading cause of it all. Hence they must have decided to emphasize the climate crisis over the biodiversity crises at the United Nations and the elaborate ruse to make it seem like fossil fuel burning is the leading cause of the climate crisis. It is more difficult for them to make the case that fossil fuel burning is the leading cause of the biodiversity crisis or any of the other nature crises we face today.

We now need put additional resources in place to bust this myth so that the climate *healing* phase can flourish on the four foundational truths of Planet B.

2. The Self Transformation:

This is the oxygen mask rule, putting on our own oxygen mask first in an emergency before helping others. If we humans are not healthy, then how can we possibly become Vitally Engaged Guardians of Animals and Nature?

When we heal ourselves, we heal the planet and conversely, when we heal the planet, we heal ourselves. This is a virtuous upward spiral and it begins with summoning the discipline to heal ourselves first.

In order to implement the 8 healing habits, we must first ensure that every human being has access to healthy whole foods plant-based vegan meals on an emergency basis. It is best to accomplish this by enrolling places of worship in a Food Healers community food distribution initiative since such places of worship reach almost every human being on the planet.

We can then help people improve their physical, mental and spiritual well-being by offering them free guidance on yoga, tai-chi, meditation, etc. I personally do ten sun salutations while breathing consciously each morning, followed by a six-minute plank pose to build my core strength. Then I meditate for 12 minutes and take my morning shower. This morning routine takes me a little over half an hour and it is sufficient to ground me for the rest of the day.

There IS a Planet B

Once again, the self-transformation movement is already well underway, even though obesity, diabetes and other chronic disease rates are also soaring across the world. Millions and millions of people are taking up yoga and meditation everywhere. Even if a many of them are exploited by a few opportunistic teachers, it shows that the intention to heal ourselves is alive and well.

3. The Goals Transformation:

Next, we must agree to transform our goals from the UN SDGs to the Vegan SDGs, dropping SDG #8 (Decent Jobs and Economic Growth) and adding SDG #18 (Zero Animal Exploitation). This requires dropping the wages for jobs mindset and seeing the nature and climate crisis for what it is, a global public health emergency that calls for an extraordinary intervention of *radical inclusion* as if our collective survival depends on it. Because it does.

Since the nations of the world have agreed to meet all the UN SDGs by 2030, we can suggest a timeline on how they can go about doing it. First, meet SDG #2 (Zero Hunger) in 2026 by making available free, healthy, starch-based, whole-foods Vegan meals to every human being on the planet in all places of worship. Then, meet SDG #3 (Good Health and Well Being) in 2027 by widely promoting the 8 healing habits. And so on.

Through an injection of systems thinking, it is not too late to actually meet these SDGs within the agreed upon timeline.

4. The Socio-Political Transformation:

The socio-political transformation that we seek are inspired by the following key questions[8]:
- How do we govern ourselves for the task of restoring the ecosystems of the planet?
- How do we ensure that every living being is free and able to contribute their gifts for accomplishing this task?

- How do we make decisions for the common good of Nature, including humanity?

Here are a few socio-political transitions that appear to be essential:
- Speciesism / Colonialism / Racism / Ableism / Patriarchy → **Veganism and Radical Sacredness**
- Individualistic → **Community oriented**
- Competitive → **Collaborative**
- Hierarchical → **Cooperative and Consensual**
- Secrecy and Lies → **Openness and Truth**
- Endless war → **Durable peace**
- Selfishness and Greed → **Selflessness and Contentment**
- Exclusivity → **Radical inclusion**
- Domination / Control relating → **Partnership / Respect relating**
- Individual transparency / Institutional privacy → **Individual privacy / Institutional transparency**
- Governance for Economic growth → **Governance for Ecological thrivability**
- Representative democracy → **Participatory democracy**
- Concentration of power → **Distribution of power**

In every one of these transitions, there are numerous organizations already working on implementing them.

5. The Economic Transformation:

The economic transformation that we seek is inspired by the following key questions[9]:
- How do we ensure that every human being has access to pure water, vitalizing food, clean air and comfortable shelter?
- How do we meet these basic human needs while ensuring that we routinely stay within the planetary boundaries?
- How do we implement processes so that robust and preventive health is normalized?

Here are a few economic transitions that appear to be essential:

There IS a Planet B

- Transaction (Illness) economy → **Gift (Wellbeing) economy**
- Endless growth → **Sustainable development**
- Debt currency → **Ecological currency**
- Fossil fuels → **Distributed renewable energy**
- AC grid → **DC grid with energy storage**
- Planned obsolescence → **Robust durability**
- Labor based on compulsion → **Labor based on volunteerism**
- Commodification of life → **Decommodification**
- Conspicuous consumption → **Conscious simplicity**
- Toxic products and processes → **Non-toxic products and processes**
- Reactive medicine → **Preventive and Lifestyle Medicine**

Once again, in every one of these transitions, there are numerous organizations working on effecting them.

6. The Ecological Transformation:

The Ecological transformation that we seek are inspired by the following questions[10]:
- How do we manage human impact on the ecosystems of the planet?
- How do we manage relationships of mutual respect and avoidance with the predators in ecosystems?
- How do we ensure that humans are routinely giving more to Nature than taking from Nature?

Here are a few ecological transitions that appear to be essential:
- Monocultures → **Biodiversity**
- Ecological overshoot → **Ecological thrivability**
- Climate *heating* → **Climate *healing***
- Artificial scarcity → **Natural abundance**
- Death dealing → **Life enriching**

In every one of these transitions as well, there are numerous organizations working on effecting them.

7. The Spiritual Transformation:

The spiritual transformation is to help us return home to our true vegan nature and ensure that generations to come will preserve the sacredness of life on earth[11]. We are mostly Born Vegan[12], but we get educated and socially conditioned out of it in the climate *heating* phase of our civilization.

The main transitions that appear to be necessary for the spiritual transformation are:
- Animal slavery → **Animal liberation**
- Deliberate cruelty → **Compassion for all life**
- Religious divisions → **Acceptance of all faith and wisdom traditions**
- Egocentric → **Ecocentric**
- Fear of Death → **Love of life**

Once again, in every one of these transitions there are numerous organizations working on realizing them on Planet B.

---

[1] The saga of the Kolkata Metro has been chronicled on Wikipedia here: https://en.wikipedia.org/wiki/Kolkata_Metro link accessed on March 8, 2025.

[2] The Delhi Metro was financed by the Japan International Cooperation Agency (JICA) and its implementation was meticulously documented in this 2016 report by Yumiko Onishi, *Breaking Ground: A Narrative on the Making of the Delhi Metro*, link https://www.jica.go.jp/Resource/activities/evaluation/ku57pq00001zf034-att/analysis_en_01.pdf accessed on March 8, 2025.

[3] The McDougall Foundation is an environmental non-profit dedicated to informing global citizens about the positive effects of dietary-therapy on chronic disease and our planet: Link https://www.mcdougallfoundation.org/ accessed on March 8, 2025.

There IS a Planet B

---

[4] Dr. McDougall's ten favorite one-liners:
   1. The fat you eat is the fat you wear
   2. Starches make you thin
   3. Sugars do not ordinarily turn into fat
   4. Sugar satisfies the hunger drive
   5. Protein deficiency is impossible, even on a vegan diet
   6. There is no such thing as dietary calcium deficiency
   7. Plants, not fish, make all Omega-3 (good) fats
   8. Taking vitamin supplements will increase cancer, heart disease, and death
   9. In order to get the cure, you must stop the cause
   10. People love to hear good news about their bad habits

For more details, please see https://www.drmcdougall.com/misc/2013nl/jul/one.htm, link accessed on March 8, 2025.

[5] Lisle, D., and Goldhamer, A., *The Pleasure Trap: Mastering the Hidden Force that Undermines Health and Happiness,* Book Publishing Company, 2017. Link https://www.healthpromoting.com/pleasure-trap-0 accessed on March 8, 2025.

[6] Watson-Tara, M., *Go Vegan: A guide to delicious everyday food – for the health your family and the planet*, Lotus Publishing, 2019. Link https://macrovegan.org/macrovegan-services/books/ accessed on March 8, 2025. Marlene's work transcends traditional and cultural boundaries to include food ingredients commonly found today.

[7] Shukla, S., *Plant Based India: Nourishing recipes rooted in tradition,* The Experiment Publishing, Aug 2022. Link https://theexperimentpublishing.com/catalogs/spring-2022/plant-based-india/ accessed on March 8, 2025. Dr. Sheil Shukla's work reinvents traditional Indian recipes in an oil-free, whole-foods, plant based vein.

[8] The socio-political transitions are examined in detail on our Vegan World 2026 web pages here: https://climatehealers.org/veganworld/political/ and https://climatehealers.org/veganworld/social/ links accessed on March 8, 2025.

[9] The economic transitions are examined in detail on Vegan World 2026 web page here:
https://climatehealers.org/veganworld/economic/ link
accessed on March 8, 2025.

[10] The ecological transitions are examined in detail on our Vegan World 2026 web page here: https://climatehealers.org/veganworld/ecological/
link accessed on March 8, 2025.

[11] The spiritual transitions are examined in detail on our Vegan World 2026 web page here:
https://climatehealers.org/veganworld/spiritual/ link
accessed on March 8, 2025.

[12] It is my privilege to acknowledge Sarina Farb who was born vegan and remained steadfastly vegan in a world where the system was constantly tempting people like her to stray. Please check out her journey here: https://www.bornvegan.org/,
link accessed on March 9, 2025.

29. The Money Game Transformation

"You can't fundamentally change big systems.
You can only abandon them and start over."
– Deborah Frieze

As humans, we tell common stories and we play common games in order to coordinate our actions among millions and billions of us. This has made us the most powerful species, even capable of significantly disrupting the life-support systems on the planet.

The story of nation states is just that, a story. The borders of nation states are quite often just lines drawn on a map by someone a long time ago and yet we sing anthems, go to war and kill each other to preserve the sanctity of these lines.

The story of money is just that, a story. Paper currency is literally just pieces of paper with pictures of dead luminaries and numeral inscriptions written on them. Yet, we've killed each other over these pieces of paper.

We play common games in which there are rules that we all agree upon and which result in real world outcomes. The most significant game we play today is, of course, the game of money. In this game, there are three fundamental roles for players: a Buyer, a Seller and a Banker.

There IS a Planet B

Alice wishes to buy a piece of property from Charles, the Seller. Alice goes to Bob, the Banker, and asks for a loan to buy the property because she doesn't have enough money in her account. According to the rules of the money game played today, Bob is allowed to create money out of thin air, provided Alice or someone else had deposited at least 10% of the loan amount with Bob for safekeeping.

Bob creates the money out of thin air and deposits it in Alice's account. Alice then transfers the money over to Charles's account and receives the deed for the property. However, as long as Alice has not paid back the loan along with the interest to Bob, the property actually belongs to Bob, the Banker, because he wrote the loan using the property as collateral.

Alice now has to extract from the property (or the planet) and create enough wealth in order to settle the loan with Bob and have some left over for her own use. Otherwise, the rights to the property transfers to Bob and Alice can even starve to death.

These are the rules of the game. We then form governments to protect the property rights of people and ensure that these rules are followed. We build prisons to incarcerate those who don't follow these rules.

Notice that in the above transaction, the default owner of the property and by extension, the default owner of the entire planet is the Banker. For every transaction needing a loan, the Banker takes over ownership of the property being transacted by creating money out of thin air.

At any given point in time in this money game, there is more debt than money. The game depends on economic growth to sustain itself, just as any Ponzi scheme depends on growth to sustain itself. The money game as constituted today is a fear-based hunting game in which we face the prospect of starvation if we don't succeed in extracting enough from the planet to return the loans and grow our net worth.

175

Also, at any given point in time in this money game, roughly one-tenth of the players are left starving so that there is a steady stream of people willing to do the kinds of work, such as mining, slaughterhouse work, etc., that no one would want to do otherwise. The game has built a transactional economy in which everyone is expected to maximize their profit. It is a game in which we routinely take more than we give, because otherwise we would soon go bankrupt and join the ranks of the hungry.

This game of money fundamentally creates all the characteristics listed on the climate *heating* side of the transitions listed in the previous chapter: Speciesism / Colonialism / Racism / Ableism / Patriarchy, Endless war, Selfishness and greed, Concentration of power, Monocultures, Ecological overshoot, Artificial scarcity, Animal slavery, Fear of death, etc.

Therefore, the civilizational transformation from the climate *heating* phase to the climate *healing* phase requires changing our money game, from this fear-based hunting game for artificially scarce resources in which we face the prospect of starvation if we don't succeed, to a new game based on love, generosity and natural abundance. The architecture and rules for this new game would incorporate characteristics that are the exact opposite of those used in the current money game.

As an aside, there are some people who believe that we don't need money at all to create a compassionate, love-based society. I agree with them provided we were dealing with a population of 8 billion enlightened human beings at the level of a Buddha or a Jesus or a Krishna. I feel that in the real world today, we need to create a new money game to replace the one that we have been playing for so long.

As Buckminster Fuller said, *"If you want to teach people a new way of thinking, don't bother trying to teach them. Instead, give them a tool, the use of which will lead to new ways of thinking."*

The new money game is the tool leading to new ways of thinking that fundamentally creates characteristics listed on the climate

There IS a Planet B

healing side of the transitions in the previous chapter: Veganism and Radical sacredness, Durable peace, Selflessness and generosity, Distribution of power, Biodiversity, Ecological thrivability, Natural abundance, Animal liberation, Love of life, etc.

In the new game, code named *Aquarius*[1], the money flows from the bottom up in the form of ecological allowances from Nature, not top down in the form of debt issued by bankers. Think of it as a Universal Basic Income (UBI) offered by Mother Earth, proportional to her overall well-being, because she loves us. Healthy food, shelter, clothing and the basic needs of humanity will be met for free so that each of us know that we belong in the family of humans and can securely dedicate ourselves to the task of restoring the ecosystems of the planet. Our purpose would be to restore the blue green planet that we inherited 10,000 years ago and the biodiversity that we lost during these 10,000 years.

In the new money game, humans belong to the Earth and are the caregivers of the Earth, but the Earth does not belong to humans. The object of the game is to meet our ecological responsibility to ensure a stable and thriving planet. Instead of governing to enforce the rights of people, we would be governing to ensure that everyone is able to meet their ecological responsibilities. Labor would be based not on compulsion, but on volunteerism. Transactions occur in the Gift economy where we would routinely give more than we take.

Cooperative work is done through crowd-sourcing the resources we need. Products are priced according to the ecological footprint for creating them. When people consume products, their footprint account is deducted by the appropriate amount and transferred to the producer's account. When a producer creates the replacement product, that footprint value is permanently retired from the producer's account.

This ensures that the ecological footprint of humanity never exceeds the limit for ecological thrivability on Planet B.

[1] The new money game of **Aquarius** helps ensure that humans don't exceed planetary boundaries while meeting the basic needs of humanity. Link https://climatehealers.org/transform/that/ accessed on March 8, 2025.

There IS a Planet B

30. How YOU Can Help

"Do everything you can possibly do, and then a lot more."
– Kevin Anderson

If you have read this far, then you are probably wondering how you can be an active participant in this greatest transformation in human history? The answer is quite simple. If you are not already active, then it is time to get active. If you are already active, then it is time to get more active.

In her **Two Loops Theory of Change**, Deborah Frieze[1] proposes that there are four kinds of roles in this transformation:
1) **Trailblazers**, the intellectuals designing Planet B;
2) **Illuminators**, the artists and story tellers who promote Planet B;
3) **Protectors**, those who resource and protect Planet B; and
4) **Hospice workers**, those who clean up and help Planet A die out and transition to Planet B.

In the Bhagavad Gita[2], the gospel of action that is part of the great Indian epic, *Mahabharata*, Lord Krishna instructs Arjuna, the chief protagonist of the epic, on how to get active in these roles, also known as *Varnas (Brahmins, Vaishyas, Kshatriyas and Shudras)*.

Mahabharata relates to the conflict between the forces of love and light, the Pandavas, and the forces of fear and darkness, the Kauravas. On the eve of the battle, Arjuna, the Pandava brother who represents courage, seemingly loses his nerve to fight and seeks counsel from Lord Krishna.

There IS a Planet B

The Bhagavad Gita is the dialogue that ensues in which Lord Krishna instructs Arjuna on why and how he must fight this quintessential battle of life.

On the question of how to get active, he states that there are four considerations:

1. *Sva Dharma,* or what is the right action based on your skills and the resources available to you,
2. *Kula Dharma,* or what is the right action based on your family circumstances,
3. *Yuga Dharma,* or what is the right action based on the era that we are living in, and
4. *Sanatana Dharma,* or what is the right action based on universal considerations of love and compassion.

You must take these four aspects of right action into consideration and then choose your role. Furthermore, once you take action, you must *let go of the fruits of your action*, evaluate the results and then repeat the same process.

Lord Krishna tells Arjuna that the right action chosen according to these considerations is always better than the right action told to us by someone else.

Finally, after addressing almost every ethical and moral question concerning the philosophy of life, Lord Krishna tells Arjuna, *"I love you so much that I set you free even from me and you are free to choose."*

A system based on love on Planet B has freedom at its core.

---

[1] Deborah Frieze's Two Loops Theory of Change is expressed succinctly in this video: https://www.youtube.com/watch?v=2jTdZSPBRRE accessed on March 7, 2025.

[2] The Bhagavad Gita is Hinduism's core sacred text, a dialogue of and on dharma, right action. It enshrines the essential values of the Vedic, Upanishadic and epic traditions - shruti (the revealed) and smriti (the remembered). Its structure is informal questions and answers; its mode is enquiry and search; its goal is self-discovery and spiritual illumination.
https://www.abebooks.com/9788174363244/Bhagavad-Gita-P-Lal-8174363246/plp, link accessed on March 7, 2025.

Epilogue: The Hero's Marathon Journey

*"The hero journey is inside of you; tear off the veils
and open the mystery of yourself."*
– Joseph Campbell

We see this greatest transformation in human history as a culmination of the hero's marathon journey for humanity. Two hundred thousand years ago, we were all born to the same mother, our mitochondrial Eve. We were all just one family in Africa when our ancestors started the climate *heating* project some 50,000 years ago. We migrated out of Africa and went to every part of the globe, armed with the technology of fire.

We suffered a lot in the process, having to cross over ice bridges, while dealing with fearsome predators and food scarcity in the harsh terrains that we migrated into. Nevertheless, we persisted because anything worthwhile rarely comes easily.

During the current interglacial, we overcame superstitions and ignorance. We developed written language, mathematics, sciences, arts and technology. In fits and starts, we bent the arc of the moral universe towards justice, because that arc does not bend by itself.

We endured unspeakable violence in wars, famines, floods and droughts. We created the conditions for the abundance we

There IS a Planet B

experience today. This abundance is now allowing us to become the caregiver species of the planet, to usher in the climate *healing* phase of our civilization.

This is as if humanity has been running a marathon race in which we have completed 26.197 out of 26.2 miles, with only 14 feet to the finish line. But to get to the finish line, we have to make a U turn to a **VEGAN** ethic and lifestyle, to stop *heating* and start *healing* the planet.

On the other hand, if we go straight instead of taking the U turn, we see a prototypical American fast food outlet, with a red-nosed clown standing outside enticing us to come in. But we know that if we take that detour from the task at hand, that clown is going to kill us, our children and our grandchildren.

What will we choose to do? What will YOU choose to do? Will you choose to put on your oxygen mask, undergo your self-transformation and join the fastest growing social justice movement in the world to help bring about the greatest transformation in human history?

The power is in your hands.

With a nod to Patagonia[1] and with gratitude to Ray Kowalchuk, on Planet B:

**Veganism is essential to a safe living space for
humanity And we don't believe anyone who says
That culture and traditions are more important than the
environment
Because the reality is
Our relationships with animals matter
So don't tell us that
We've run out of solutions**

(On Planet A, we have been reading this from the bottom up.)

[1] Patagonia penned a reversible ad about the nature and climate crisis in 2020:

We are all screwed
So don't tell us that
We can imagine a healthy future
Because the reality is
It's too late to fix the climate crisis
And we don't trust anyone who says
We need to demand a livable planet
Because we don't have a choice
(Now read this bottom up)

Ray Kowalchuk modified this ad to describe the Planet B vs. Planet A conundrum. Link to an article on the ad https://www.thedrum.com/news/2020/11/30/patagonia-pens-reversible-poem-raise-awareness-climate-crisis accessed on March 8, 2025.